How to Repair
Briggs & Stratton
Engines

About the Author

Paul Dempsey is a master mechanic and the author of more than 20 best-selling technical books, including *Small Gas Engine Repair, Second Edition* and *Troubleshooting and Repairing Diesel Engines, Third Edition,* both available from McGraw-Hill. He has also written over 100 magazine and journal articles on topics ranging from teaching techniques to maintenance management to petroleum-related issues.

How to Repair Briggs & Stratton Engines

fourth edition

Paul Dempsey

New York Chicago San Francisco Lisbon London Madrid
Mexico City Milan New Delhi San Juan Seoul
Singapore Sydney Toronto

The McGraw·Hill Companies

Library of Congress Cataloging-in-Publication Data.

Dempsey, Paul.
   How to repair Briggs & Stratton engines / Paul Dempsey. — 4th ed.
      p.   cm.
   Includes index.
   ISBN 0-07-149325-5 (alk. paper)
      1. Internal combustion engines, Spark ignition—Maintenance and repair.
   I. Title.   II. Title: How to repair Briggs and Stratton engines.
   TJ789.D43   2007
   621.43'40288—dc22

                                                            2007007217

McGraw-Hill books are available at special quantity discounts to use as premiums and sales promotions, or for use in corporate training programs. For more information, please write to the Director of Special Sales, Professional Publishing, McGraw-Hill, Two Penn Plaza, New York, NY 10121-2298. Or contact your local bookstore.

1 2 3 4 5 6 7 8 9 0   DOC/DOC   0 1 2 1 0 9 8 7

ISBN 13: 978-0-07-149325-3
ISBN 10: 0-07-149325-5

This book is printed on acid-free paper.

Sponsoring Editor Larry S. Hager	**Indexer** Kevin Broccoli
Editing Supervisor David E. Fogarty	**Production Supervisor** Richard C. Ruzycka
Project Manager Madhu Bhardwaj	**Composition** International Typesetting and Composition
Copy Editor Raina Trivedi	**Art Director, Cover** Jeff Weeks
Proofreader Ragini Pandey	**Cover Designer** Malvina D'Alterio

Contents

Introduction   ix

Safety considerations   xi

1 • The product range   *1*

Engine identification   2
Buying an engine   4
Buying part   7
Operating cycle   8
Nomenclature & construction   10
Displacement, hp, torque,
    and rpm   17

2 • Troubleshooting   *21*

Failure to start   24
Loss of compression   27
Engine runs 3 or 4 minutes and
    quits   29
Loss of power   29
Failure to idle   30
Failure to shutdown   30
Kickback   30
Starter binds   30
Exhaust smoke   31
Excessive vibration   31

3 • Ignition systems *33*

Spark plugs 33
Flywheels 34
Magnetos 39
Armature air gap 45
Magnetron 46
Shutdown switches 53
Safety interlocks 53

4 • The fuel system *57*

How carburetors work 57
External adjustments 62
Troubleshooting 64
Removal & installation 67
Repair & cleaning 71
Carburetor service by model 72
Fuel pump 82
Governors 84

5 • Starters *87*

Rewind starters 87
Eaton-pattern, side-pull starter 88
Briggs & Stratton starters 96

6 • Charging systems *109*

Storage batteries 109
Alternators 111
The Nicad system 121

7 • Engine mechanics *123*

Resources 123
Diagnosis 125
Cylinder head 129
Valves 130
Flange 139
Camshaft 146
Connecting rods 153
Piston 158
Rings 162
Cylinder bore 166
Main bearings 172
Magneto-side seal 173
Crankshaft 174

8 • The Europa *177*

Technical description   178
A closer look   179
Service history   193

9 • Winds of change *197*

Clean air   199

Index   207

Introduction

This fourth revision of *How to Repair Briggs & Stratton Engines* is an attempt to take the mystery out of small-engine repair. Small engines are simple devices, on a par with automobiles of the 1930s. Anyone blessed with a bit of mechanical aptitude and the patience to see things through can do the work. In addition to the money saved—some shops extract a $60 minimum labor charge for adjusting a throttle cable—there is a joy in making an inert collection of aluminum and iron come alive. They say all working people remember their first paycheck, the pride they experienced in being worth something to the larger world. The glow that comes when you get your first engine started is equally unforgettable.

This book covers all American-made Briggs single-cylinder engines built during the last 30 years or so. The emphasis is on the newer models, but things you need to know when you encounter an old timer are also included.

Unlike factory shop manuals, this book is intended for the amateur or beginning mechanic. The first chapter covers the basics, how engines work and what terms like "torque" and "compression ratio" mean. The next chapter goes into detail about troubleshooting, which is the essential skill of the mechanic. Material included here enables the reader to isolate the problem to the system involved. Once you have located the problem in, say, the fuel system, turn to that chapter. There you will find information on how to track the malfunction down to a particular component and instructions for making the repair. Each system—fuel, ignition, rewind and electric starters, alternators, and engine mechanics—has its own chapter.

The "Engine Mechanics" chapter describes the repair and restoration of major engine components. Here you will learn how to reface a cylinder head and other warped castings with a homemade "Armstrong milling machine," the meaning of piston wear patterns, and how to do everything necessary to bring a worn engine back to as-new condition. Some repairs, such as grinding the crankshaft and reboring cylinders, are beyond the scope of the home mechanic and must be farmed out. These processes are described in enough detail so that the reader can verify that the work was performed correctly.

Briggs supplies dealer mechanics with a range of special tools, which can be purchased by the public. But the tools are not cheap and there are often delays while you wait for delivery. Whenever possible, the text suggests substitutes. Equivalents for many of these tools can be fabricated at home.

Each critical operation is described and illustrated with photographs and drawings. Most of these illustrations originate with Briggs & Stratton, but when Kohler, Yanmar, or some other manufacturer had a better, more informative illustration, it was used.

The next chapter provides a detailed analysis of the Europa, the first modern Briggs & Stratton overhead engine and the pattern for those that followed. The engine was purchased new, disassembled and each part examined and photographed. Then it was run for more than 500 hours as a test of durability. As far as I know, this is the only published account of a long-term, real-world test of a small engine.

The book concludes with a brief look at the future. Briggs & Stratton faces challenges from foreign competition and from United States and European moves to limit exhaust emissions. This chapter describes the actions Briggs is taking to retain its position as the leading small engine manufacturer in the world.

Safety considerations

Hazards encountered in the normal course of repair/maintenance work are flagged in the text by **Warnings** and **Cautions**. **Warnings** means risk of personal injury; **Caution** means risk of equipment damage. Only the most obvious hazards are called out. I cannot anticipate all the ways of getting hurt or of damaging engines. You should always read the repair steps in their entirety before you begin any procedure.

Safety depends on the attitude of the mechanic and, to no less a degree, on the environment in which he works. A disorganized, untidy shop invites accidents.

Do:

- Keep a fire extinguisher ready when servicing small engines.
- Handle gasoline with extreme caution. Refueling and any service operation that can result in spilled gasoline should be performed outdoors, away from possible ignition sources.
- Disconnect and restrain the spark-plug wire before working on the underside of rotary lawnmowers and other engine-driven equipment.
- Wear eye protection when using a grinder, power-driven wire brush, or hammer.

Do not:

- Refuel an engine while it is running or still hot. Allow at least five minutes for the engine to cool before refueling.

- Attempt to clear a flooded engine by removing the spark plug and cranking. If compressed air is not available, install a fresh spark plug and crank with the throttle fully open.
- Operate an engine in an enclosed area. The exhaust contains carbon monoxide, a lethal gas.
- Operate an engine without a serviceable muffler or air filter. A spark arrester must be fitted for operation in wilderness areas.
- Tamper with governor settings, bend actuating links, or substitute governor springs. Excessive no-load speed might result in connecting-rod failure or flywheel explosion.

How to Repair
Briggs & Stratton
Engines

1

The product range

Briggs & Stratton is the world's largest producer of small engines with an annual output of 10.5 million units or 30,000 a day. More than 60% of small engines sold worldwide and half of the portable generators carry the black-and-white Briggs logo. Like Harley-Davidson, also based in Milwaukee, Briggs and Stratton is an American icon.

One reason why Briggs & Stratton enjoys an almost mythic stature among Americans of a certain age is the belief that its products do not change. Wars and presidents come and go, but the Briggs side-valve lives forever. And when the mower won't start, you can repair it without a degree in engineering. The skills learned in our youth still have relevance and can be passed on to our grandchildren, that is, if they are open to a world larger and more real than computer games.

Actually, Briggs engines—even the old side-valve with its rectangular cylinder head and quaint rewind starter—have undergone more-or-less continuous change. And more changes are coming to the products and to the way the company does business.

But, withal, there is a kind of elegant simplicity to Briggs engineering. A case in point is the Easy-Spin compression release that automatically raises the intake valve without adding as single part to the engine. Another example is the nylon Pulsa-Jet carburetor with its built-in fuel pump and an almost zero requirement for machine work. Briggs was the first maker of utility engines to offer overhead valves, the first to build vertical-crankshaft aluminum blocks, and the first to use aluminum cylinder bores.

To encourage proper maintenance, spark plugs, air filters and oil-drain plugs are readily accessible. Head bolts can be reached by removing the shroud, an operation that can be accomplished in five minutes or less. Each engine undergoes a test run before shipping to determine that it starts promptly and operates within governed speeds. Most engines are sold to original equipment manufacturers (OEMs). To eliminate problems that might occur because of mounting provisions or shrouding, each of these applications must meet rigorous standards for noise, vibration, cooling, and ease of starting. In addition, the company operates the Florida Test Base at Fort Pierce where engines and the equipment they power undergo field testing.

Until recently production came from six plants located in Alabama, Georgia, Kentucky, Missouri, and Wisconsin. When you call the company with a technical problem, you speak with someone in Milwaukee.

Engine identification

Briggs currently builds more than 100 models of one- and two-cylinder engines ranging from 2½ to 31 hp, with literally thousands of detailed variations. To purchase parts you need to know exactly what engine you have. The model, type, and code numbers are on the blower shroud near the spark plug or above the muffler. In addition, you may see a "family" number, which shows that the engine has been certified by the EPA and/or the state of California. "B-1" engines cannot be sold in California because of exhaust-emissions regulations.

The model number has five or six digits (Table 1-1). The first one or two digits indicate the cylinder displacement in cubic inches rounded up to the next whole number. If the number begins with "9," you have a nine-cubic inch engine. If "11," the engine displaces 11 cid. The other numbers refer to lay of the crankshaft (vertical or horizontal), and to the type of carburetor, starter, and alternator fitted.

The type number identifies the original-equipment purchaser, such as the lawnmower-maker MTD, and any special features the purchaser required. The code number shows the date of manufacture and where the engine was built. For example, 06021504 translates as:

06 = 2006
02 = February
15 = day of the month
04 = plant number

Running production changes mean that the build date often must be supplied to obtain the correct parts.

This book covers repair procedures for single-cylinder, side- and overhead-valve engines produced in this country. Japanese Vanguard engines carry the Briggs logo and share parts with the home-grown product, but come out of a different tradition.

Table 1-1
Briggs & Stratton model number code

1st one or two digits	2nd or 3rd digit	3rd or 4th digit	4th or 5th digit	5th or 6th digit
Displacement (cubic inches)	Internal factory code that refers to cylinder type, layout of components, etc.	Crankshaft orientation	Main bearing, oiling system, and pto type	Starter and alternator type
6	0	0–4 horizontal	0—block-metal or DU bushing	0—w/o starter
8	1	5–9 vertical	1—block-metal bearing	1—rope start
9	2	A–G horizontal	2—renewable sleeve bearing, splash lubrication	2—rewind starter
10	3	H–Z vertical	3—ball bearing, splash lubrication	3—110 V starter
11	4		4—ball bearing, oil pump	4—110 V starter w/alternator
12	5		5—block-metal bearing, 6:1 gear reduction	5—12 V starter
13	6		6—block-metal bearing, 2:1 gear reduction	6—alternator w/o electric starter
16	7		7—block-metal bearing, oil pump	7—12 V starter w/alternator
18	8		8—block-metal bearing with cam-driven pto 90° to crank	8—vertical-pull rewind
19	9		9—block-metal bearing, with pto parallel to crank	9—other starter type (e.g., pneumatic)
20 and so on	A–Z		A—block-metal bearing with oil pump and no filter	A—starter motor, alternator and 120 V inverter

Buying an engine

Briggs & Stratton engines run the gamut from world-class industrial plants to discount-house specials (more than half of the company's sales are to Wal-Mart, Lowe's, and Sam's Club). Segments of the lawn-and-garden market have also experienced a race to the bottom. During the 1970s, riding mowers cost around $2400; today you can buy one for half that. Something more than hydrostatic transmissions had to give.

All Briggs & Stratton engines are built to the same basic level of precision, as determined by the tooling. That is, cylinder bores and other critical parts are held to ± 0.001 in. and crankshaft journals to ± 0.0005 in. Connecting rods are diamond-bored for surface finish, as are the cylinders on some models. From this point forward, quality varies.

If we define quality as durability, an engine buyer should look for:

- Cast-iron cylinder sleeves. These Dura-Bore cylinders wear better than the aluminum (Kool Bore) alternative and stiffen the block for more consistent power and less oil consumption.
- Overhead valves. While valve location has nothing intrinsically to do with quality, OHV engines are more recent designs that, for the most part, have other desirable features such as oil filters, low-friction piston rings, and automotive-type carburetors.
- Large air filters. The best of these are Donaldson filters used on some commercial models, but any of the replaceable paper-element filters are superior to the polyurethane filters on low-end engines.
- DU™ or ball-bearing mains. See the "Bearings" section for additional information.
- Fuel pumps. Millions of Briggs engines operate happily with gravity feed or with suction-lift carburetors. A pulse-type pump develops 1.5-psi delivery pressure regardless of the level of fuel in the tank, and enables a 75-micron filter to be used, as opposed to a 150-micron filter for gravity feed and a wide-mesh screen for suction lift.
- Centrifugal governors. These governors vary in sophistication, but all have better precision and reliability than the air-vane governors on cheaper models.
- Bolt-on mufflers. Briggs offers a pipe-threaded and noisy muffler and two bolt-ons known as the Low-Tone and Super Low-Tone. Opt for the latter, if possible.
- Inconel valves. The presence of these valves means that basic components should be good for more than 2000 hours.

Model names change with bewildering frequency, but Japanese-built Vanguards stand at the top of the line. These engines have stiffer crankcases and benefit from selective assembly. Cylinders and pistons are graded by where they fall within the tolerance spread and assembled as matched pairs.

Second in terms of quality are the American-made Intecs, which are of fairly recent design and far more sophisticated than the traditional product. That said, the 11-cid vertical-shaft Intec comes with an aluminum bore, which is hardly a recipe for longevity. Other models are even more of a mixed bag and should be scrutinized closely for the features you consider important. Quantums, 650s, and 850s are entry-level products, okay for casual use.

Your area distributor (listed in Table 1-2) can provide you with a copy of the current *Engine Sales Replacement Specifications and Price List*, which contains the performance, pricing, and dimensional data.

Critical dimensions for any application are the crankshaft diameter on the power takeoff (pto) side, how far it extends outside of the block, and what provision it has for mounting a blade or other driven element. For example, the horizontal 7-hp 1700400 comes with any of five crankshafts. Power takeoff reduction gear boxes have a 6:1 ratio, but the direction of rotation varies.

It's always attractive to replace an existing engine with a larger, more powerful unit. A larger engine should get the job done quicker and live longer in the bargain. How much power is enough is a matter of personal preference. The engine that came with the equipment met Briggs power standards, which state that 70% throttle should be enough to cope with the maximum anticipated load. This throttle angle gives acceptable engine life while, at the same time, provides a cushion of reserve power. Another way company engineers check for adequate power is to operate the equipment under full load and manually open the throttle to its stop. A speed increase of between 100 and 200 rpm suggests that engine selection is correct. It should be mentioned that holding the throttle open past the governor setting has potentially explosive effects. If you're going to do this at home, use an accurately calibrated tachometer and do not exceed 4000 rpm.

For horizontal engines, you need to know the vertical distance from the centerline of the pto stub to the engine-mounting surface. All horizontal-shaft engines have a rectangular footprint with a mounting bolt in each corner. Naturally, the footprint and crankshaft height and diameter become larger as engine displacement increases. You can often redrill mounting holes, but crankshaft height can be a problem. When possible, mount drive pulleys well inboard on the crankshaft extension to reduce side loads. In no case should the pulley or sprocket overhang the crankshaft stub.

The footprint of vertical-shaft engines—the diameter of the flange and the location of the mounting holes—varies with engine size. The length of the crankshaft extension is critical for rotary lawnmowers. A short crankshaft buries the blade under the deck, so that the operator must bulldoze the

Table 1-2

City	State	Zip	Company	Address	Phone
Aiea (Honolulu)	HI	96701	Small Engine Clinic, Inc.	98–019 Kam Highway	808-488-0711
Ashland (Richmond)	VA	23005	Colonial Power Div. of RBI Corporation	101 Cedar Ridge Drive	804-550-2210
Billings	MT	59101	Original Equipment, Inc.	905 Second Avenue North	406-245-3081
Burlingame (San Francisco)	CA	94010	Pacific Western Power	1565 Adrian Road	415-692-3254
Carpinteria (Santa Barbara)	CA	93013	Power Equipment Company	1045 Cindy Lane	805-684-6637
Charlotte	NC	28206	AEA, Inc.	700 W. 28th Street	704-377-6991
Columbus	OH	43228	Central Power Systems	2555 International Street	614-876-3533
Dallas	TX	75207	Grayson Company, Inc.	1234 Motor Street	214-630-3272
Denver	CO	80223	Pacific Power Equipment Company	1441 W. Bayaud Avenue #4	303-744-7891
Elmhurst (Chicago)	IL	60126	Midwest Engine Warehouse	515 Romans Road	708-833-1200
Foxboro	MA	02035	Atlantic Power, Inc.	77 Green Street	508-543-6911
Houston	TX	77040	Engine Warehouse, Inc.	7415 Empire Central Drive	713-937-4000
Kenner (New Orleans)	LA	70062	Delta Power Equipment Company	755 E. Airline Highway	504-465-9222
Louisville	KY	40299	Commonwealth Engine, Inc.	11421 Electron Drive	502-267-7883
Memphis	TN	38116	Automotive Electric Corporation	3250 Millbranch Road	901-345-0300
Milwaukee	WI	53209	Wisconsin Magneto, Inc.	4727 N. Teutonia Avenue	414-445-2800
Minneapolis	MN	55432	Wisconsin Magneto, Inc.	8010 Ranchers Road	612-780-5585
Norcross (Atlanta)	GA	30093	Sedco, Inc.	4305 Steve Reynolds Blvd.	404-925-4706
Oklahoma City	OK	73108	Engine Warehouse, Inc.	4200 Highline Blvd.	405-946-7800
Omaha	NE	68127	Midwest Engine Warehouse of Omaha	7706 "I" Plaza	402-339-4700
Phoenix	AZ	85009	Power Equipment Company	#7 North 43rd Avenue	602-272-3936
Pittsburgh	PA	15233	Three Rivers Engine Distributors	1411 Beaver Avenue	412-321-4111
St. Louis	MO	63103	Diamond Engine Sales	3134 Washington Avenue	314-652-2202
Salt Lake City	UT	84101	Frank Edwards Company	1284 South 500 West	801-972-0128
Somerset (New Brunswick)	NJ	08873	Atlantic Power, Inc.	650 Howard Avenue	908-356-8400
Tampa	FL	33606	Spencer Engine, Inc.	1114 W. Cass Street	813-253-6035
Tualatin (Portland)	OR	97062	Brown & Wiser, Inc.	9991 S. W. Avery Street	503-692-0330

CANADA

City	State	Zip	Company	Address	Phone
Mississauga (Toronto)	ON	L5T 2J3	Briggs & Stratton Canada Inc.	301 Ambassador Drive	416-795-2632
Delta (Vancouver)	BC	V3M 6K2	Briggs & Stratton Canada Inc.	1360 Cliveden Avenue	604-520-1294

6

mower through high grass. One that is too long flings debris out from under the machine and presents a major hazard. Briggs can furnish spacers to permit Intec engines to be fitted in place of cheaper, short-shaft models.

Part of the price of the engine is reflected in the costs of the two-year warranty for (depending upon engine model) consumer or commercial use. Emissions-related equipment is covered for the life of the engine. Briggs also has an informal, "good-will policy" that, depending upon the circumstances, can pay for repairs after the warranty period has expired.

As part of the factory's Repower Project, Briggs dealers can now supply Series 650 vertical-shaft engines, complete with a large-diameter rewind starter and Lo-Tone muffler, for $139.95, which is a bargain. The Series 850 with an automatic choke goes for $15 more. Going up the power and quality scale, the vertical-shaft 31-cid I/C 17.5, replete with overhead valves, iron cylinder liner and alternator, costs $420.

Liquidators, such as Kansas City Small Engines, are another source. These outfits purchase surplus engines from OEMs who have overestimated demand for their products. The Kansas City operation is said to do a four-million-dollar business annually on the Internet, and carries as many as 40,000 units in stock. But shop around and make sure the engine you select has the features you want and is covered by factory warranty. Granger, Northern Tool, and other industrial supply houses also sell under list, but selection is limited. Contrary to Internet rumor, Briggs does not sell seconds, i.e., engines that have failed to pass final inspection. What one sometimes sees advertised on the net as "seconds" are engines that have suffered shipping damage, such as dented shrouds or cracked spark plugs.

Buying parts

The company web site www.briggsandstratton.com provides exploded views and parts lists with prices for all current and most older engines. The availability of parts for long-obsolete engines is one reason why many of us harbor good feelings toward the company. Try to buy parts for 30-year-old Yamaha motor bike.

Carburetor, ignition, and other high-demand parts are available on the net and at auto-parts houses at better-than-list prices. But be careful: after-market parts can save pennies and cost dollars. Briggs has found polyurethane air filters that pass dust and paper elements with open pleats, like the pages of a book. After-market oil filters are no better. The paper media is of unknown performance, anti-drain back valves are liable to fail, and the paper end caps collapse, resulting in internal hemorrhage. Imported starter-motor armatures exhibit the same dismal level of quality.

Operating cycle

Four-cycle engines of all types share the same operating cycle first com-
mercialized by Nicolaus Otto back in 1876. The Otto cycle consists of four
events:

- intake of air and fuel;
- compression of the air-fuel mixture by the piston;
- ignition and subsequent pressure rise in the cylinder; and
- exhaust of the spent gases.

Four-stroke-cycle engines require four strokes of the piston, or two full
revolutions of the crankshaft, to complete the cycle. Figure 1-1 illustrates the
sequence of piston and valve movement. The cycle begins with the piston
moving downward in the bore with the intake valve open. Air and fuel,
impelled by atmospheric pressure, enter the cylinder. The exhaust valve
remains closed during the intake and subsequent compression stroke. As
the piston reaches the lower limit of its travel, the intake valve closes.

The piston then rounds bottom dead center (bdc) and begins its climb up
the bore. Since both valves are closed, the air-fuel mixture is trapped
between the top of the piston and the underside of the cylinder head.
Compression ratios range from about 6:1 for side-valve engines to more than
8:1 for overhead-valve models. All things equal, the higher the compression
ratio, the greater the power output.

As the piston approaches top dead center (tdc), the spark plug fires to
ignite the charge. The force of the explosion drives the piston down on the
power stroke. The flywheel accelerates, absorbing energy that will be
returned to the system during the intake, compression, and exhaust strokes.

The exhaust valve opens near the end of the power stoke. Spent gases
spill into the atmosphere blown down by residual cylinder pressure and
subsequently by the piston as it climbs back toward tdc. The exhaust valve
closes near tdc and the intake valve opens to initiate another cycle.

Note that events do not perfectly correlate with piston strokes. The intake
valve stays open late, a few crankshaft degrees into the compression stroke.
The incoming charge has inertia and streams in against compression to better fill the cylinder. In the like manner, the exhaust valve opens early before
the piston reaches bdc on the power stroke. Time is required to overcome
inertia and get the exhaust gases moving around the open valve.

While we're on the subject of valves and timing, it's interesting to note that
engines are timed from what mechanics call the "rock position." You might
try this experiment when you have an engine apart. Turn the crankshaft to
bring the piston to tdc. Now, without looking at the timing marks, install the
cam. Reposition the camshaft until one or two degrees of crankshaft movement just cracks the intake valve open, and the same amount of movement

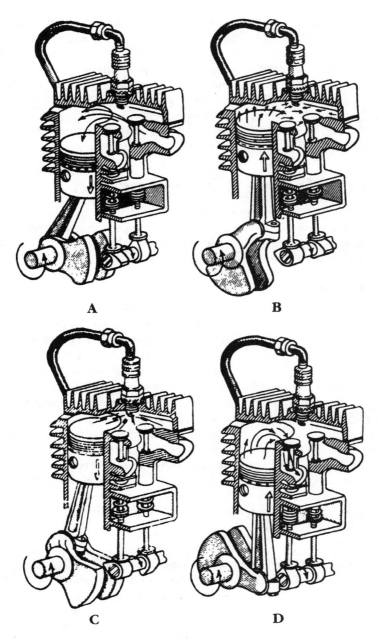

A   B

C   D

FIG. 1-1. *The four-stroke cycle: (A) intake; (B) compression; (C) power or expansion; (D) exhaust. Tecumseh Products Co.*

in the other direction does the same for the exhaust valve. This is the rock position. Now look at the timing marks. They will be in alignment.

Timing from the rock position has saved many a mechanic's bacon when the crankshaft timing mark, as sometimes happens, has worn away. It works for engines of all types with conventional, nonracing cam grinds, but the spring-loaded camshaft flyweights used on some engines complicate things. These flyweights hold the exhaust valve open a few thousandths of an inch during starting. In order to get an accurate fix on exhaust valve movement, you must temporarily secure the weights in the running position.

Nomenclature & construction

Figure 1-2 identifies the major components that make up a typical single-cylinder, air-cooled, side-valve engine. The example illustrated has ball-type main bearings, splash lubrication, and a centrifugal governor. These features are fairly unique to small engines.

Valve mechanism

Side-valve, L-head, or flathead engines—the names all mean the same—have valves inside the block, alongside of the piston (Fig. 1-3). Side-valve engines have the virtue of simplicity, but their days are numbered. The location of the valves dictates the shape of the combustion chamber, which must be off-set to one side of the bore. The chamber has a large surface area relative to its volume. These surfaces bleed off heat that should go into the work of driving the piston. In addition, contact with the cold metal dampens the flame and increases the level of unburned hydrocarbons in the exhaust.

Overhead-valve, I-head, or valve-in-head engines locate the valves above the piston. Pushrods and rocker arms transfer motion from the camshaft as shown in Fig. 1-4. This configuration results in a symmetrical combustion chamber centered over the bore. A chamber of this type has relatively little surface area relative to its volume, which results in cleaner and more efficient combustion. The short flame path permits higher compression ratios.

Air cooling

The flywheel doubles as a centrifugal fan. Blades on the rim of the flywheel draw air from the hub area and discharge into the shroud, or blower housing. Fins cast into the cylinder head and barrel transfer about a third of the heat developed in the engine to the atmosphere. Some newer engines, such as the vertical-shaft Europa, have fins on the underside of the crankcase in the air blast generated by the mower blade. The effectiveness of sump cooling is debatable, since external fins don't do much to lower oil temperature and, in any event, the fins soon clog. But the idea is intriguing.

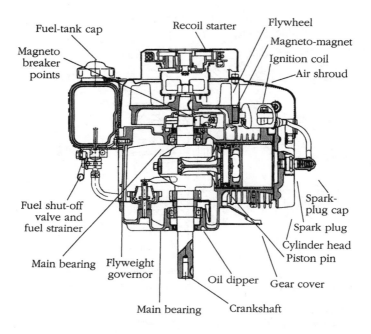

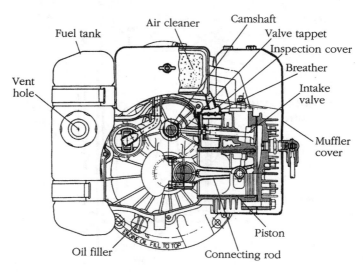

FIG. 1-2. *Small engines generally follow the same formula with only subtle differences between the most and least expensive.* Wisconsin Robin W1-145 V shown.

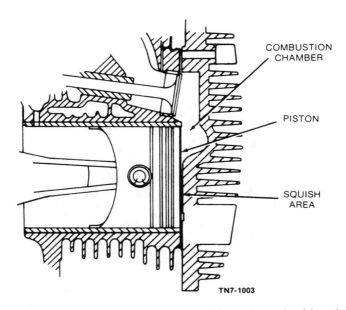

COMBUSTION CHAMBER

PISTON

SQUISH AREA

TN7-1003

FIG. 1-3. *L-head engines house their valves in the block, alongside of the cylinder bore. The relation between the valve heads and bore centerline resembles the letter "L." Onan.*

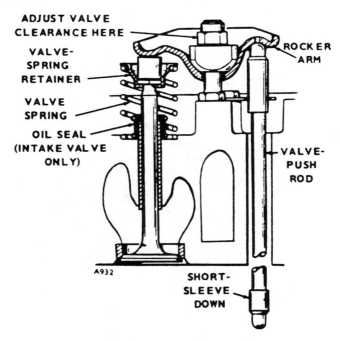

ADJUST VALVE CLEARANCE HERE

ROCKER ARM

VALVE-SPRING RETAINER

VALVE SPRING

OIL SEAL (INTAKE VALVE ONLY)

VALVE-PUSH ROD

A932

SHORT-SLEEVE DOWN

FIG. 1-4. *The overhead-valve mechanism on US-built Briggs engines follows automotive practice, in that stamped steel rocker arms ride on adjustable fulcrums. The seal prevents oil from being drawn into combustion chamber during coast down. Onan.*

12

Air cooling is the practical choice for small utility engines. The major drawback is the difficulty of controlling engine temperature with a centrifugal fan that turns crankshaft speed and rapidly loses efficiency as rpm drops. Engines overheat when bogged under load or when idled immediately after hard use. About all the owner can do is run the engine at a comfortable speed and periodically clean the cooling fins (Fig. 1-5).

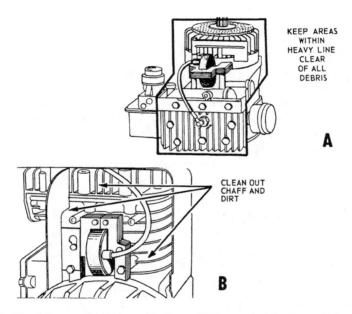

FIG. 1-5. *Head fins are critical, especially on OHV engines where as much as 80% of the thermal load passes through the cylinder head.*

As a point of interest, Briggs temperature guidelines are to run the engine under 75% rated load (corresponding to half throttle) for consumer equipment and at full load for commercial equipment. Measurements are taken when oil temperature shows no increase over a five-minute period. Maximum allowable temperature increases *over ambient* are:

- Cooling air (measured at the shroud discharge)   45°F
- Fuel in tank   30°F
- Lube oil
  Synchro-Balanced & Twin Cylinder   210°F
  Non-Synchro Balanced   190°F
- Cylinder head (measured at spark-plug gasket)   430°F

Owners, especially owners of generator, blower, and other hardworking engines, can find it useful to monitor the oil temperature. This can be easily done on engines with dipsticks. Shut the engine off, and insert a thermometer into the dipstick boss to a depth of about 5/16 in. past the end of the stick. Do not place the thermometer near or against crankcase metal, as this will result in a low reading.

Cylinder-head temperature measurements require a thermocouple integral with the spark-plug gasket, which can be purchased from a karting supply house.

Lubrication

Most Briggs vertical-crankshaft engines use splash lubrication generated by a camshaft-driven impellor known as a slinger (Fig. 1-6). Horizontal-crank models have a dipper on the end of the connecting rod. Splash lubrication works fine for side-valve engines with the mechanicals confined inside of the crankcase. Overhead-valve engines incorporate a pump to deliver oil under pressure to the valves, which can also be used to lubricate the cam and crankshaft journals. Several have replaceable oil filters.

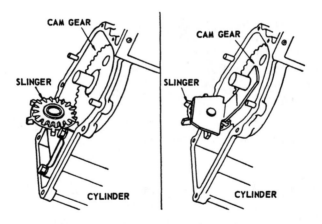

FIG. 1-6. *Slingers used on vertical-shaft engines drive off the camshaft.*

Crankcase and block

Briggs & Stratton engines made in this country have aluminum blocks. The switch from cast-iron was initiated in 1953 and within a few years annual production of light-metal engines exceeded two million units. Aluminum weighs about a quarter as much as cast iron and transfers heat nearly three times as rapidly. The material invites precision die casting techniques, which eliminate much machine work and permit closely spaced, deep cooling fins. The old iron heads are almost bald by comparison.

The better engines have cast-iron cylinder liners; light- and one or two medium-duty engines use the block metal itself as the bore. To prevent scuffing, the pistons of these Kool-Bore engines are chrome-plated, but the iron rings bear directly against the raw aluminum cylinder walls. Ryobi, Tecumseh, and other small-engine manufacturers do the same. Makers of more durable goods, such as touring motorcycles and Porsche automobiles, also employ aluminum bores, but treat the metal for wear resistance.

The company argues that Kool-Bore engines, which are primarily intended for walk-behind mowers, are adequate for the task. Studies show that the typical owner runs his or her mower for about 30 hours a year and scraps the machine after 7 years, or after 180 hours or so of operation. Briggs products undergo a minimum of 200 hours of field testing at the Florida facility, which should assure that Kool Bores live up to customer expectations. And when the engines wear out, they can be bored 0.020-in. oversize.

Although out of fashion now, iron blocks have advantages. If you run across an old cast-iron 5S, 6S, or the like in a garage sale, buy it. These engines are as durable as rocks and eminently repairable. The factory still stocks most of the basic parts, although magneto components are becoming scarce and expensive when found.

The big advantage iron has over aluminum is rigidity. Once an iron block has been seasoned by repeated heating and cooling cycles, its dimensions remain constant. Crankshafts run true and cylinder bores remain round. When an iron block is combined with very conservative power outputs, we have an engine that lives almost forever, as demonstrated by the one-lung pump engines used in the oil field. Some go for 50 years and more with only oil and spark-plug changes.

Bearings

Most engines run their crankshafts against the aluminum block and flange without an intervening bearing. The high-silicone aluminum alloy makes a reasonably durable bearing so long as the oil remains clean. Better engines employ replaceable DU bushings. These Teflon-impregnated bronze bushings tolerate twice the radial loads of aluminum bearings (Table 1-3), and have inherent lubricity, which is a useful quality for the magneto-side bearing on vertical-shaft engines without forced lubrication.

Ball bearings are expensive, noisy, and probably no more durable than DU bushings. But unlike bushings, ball bearings tolerate significant amounts of axial load, or thrust. However, the ability to handle thrust is unidirectional. As installed, bearing thrust faces are inboard. On vertical-shaft engines, the

Table 1-3
Maximum allowable main-bearing loads by type

Bearing type	Radial, or side load acting on the end of the shaft	Thrust, or axial load
Aluminum pto and magneto	200 lb	50 lb in either direction
Ball-bearing pto with aluminum or DU magneto	400 lb	50 lb inward (toward block) 500 lb outward
Ball-bearing pto and magneto	400 lb	500 lb in either direction
DU pto and aluminum or DU magneto	400 lb	50 lb in either direction

pto bearing carries downward thrust and the magneto-side bearing upward thrust.

Suppose you want to drive a centrifugal pump directly off the crankshaft, without the intervention of a belt. The thrust developed by most pumps tends to pull the crankshaft away from the block, which means that a ball bearing should be present on the pto end of the crank. On the other hand, a high-lift mower blade can generate enough thrust to require a magneto-side ball bearing. When ball bearings are fitted at both ends of the crank, the direction of thrust is no longer a matter of concern.

Aluminum connecting rods do not have replaceable inserts, but automotive machinists can regrind crankshafts to accept 0.020-in. undersized rods.

Flywheel

The flywheel acts as a reservoir to smooth power impulses and store energy that is released when the engine slows under load. Vertical-shaft, direct-drive lawnmowers receive most of the inertia necessary for starting from the blade and can use a lightweight aluminum flywheel. But if the engine drives through a belt or clutch, a heavier (and more expensive) wheel is required. The standard flywheel for 9- and 11-cid (cubic inch displacement) engines develops 9.8 lb/in.^{2} of inertia, while the heavy wheel produces almost three times the inertia at 26.7 lb/in.^{2}

A taper and nut secure the flywheel to the crankshaft. The soft aluminum key is merely a reference to establish correct ignition timing. In an emergency, say, when the key has sheared and no replacement is immediately available, you can align the keyways by eye.

Magnets cast into the rim of the wheel generate the field for the ignition system and, when fitted, for the alternator. One has to be careful when swapping flywheels between engine models. While many flywheels physically inter-change, the position and number of magnets varies with the application.

Carburetion

As mentioned previously, the better engines employ automotive, float-type car-buretors with generously sized air filters. Low-end models use diaphragm-type "mixing valves" mounted over the fuel tank. Current versions employ either an automatic choke or a primer pump to enrich the mixture for starting. These carburetors are often coupled to a polyurethane air filter that requires frequent cleaning and re-oiling.

Ignition

All modern Briggs engines employ a solid-state pulse generator known as the Magnetron. Unlike old-style magnetos with their vulnerable contact points, the Magnetron has no moving parts and rarely fails. But spark plugs remain a source of frustration for small-engine owners.

Displacement, hp, torque, and rpm

It is no accident that the first set of digits in the Briggs & Stratton model code expresses displacement in cubic inches. Displacement—the cylinder vol-ume swept by the piston—is the ground-zero measurement of engine potential, in the same way that square footage is the single most important variable in the design of a house.

The formula for calculating displacement is:

bore $\times$ bore $\times$ number of cylinders $\times$ stroke $\times$ 0.7858 = displacement

If bore and stroke are expressed in inches, the formula gives displace-ment in cubic inches, which is expressed as cid. The Sprint 96900 has a 2.56-in. bore and a 1.75-in. stroke:

2.56 in. $\times$ 2.56 in. $\times$ 1 cylinder $\times$ 1.75 in. $\times$ 0.7858 = 9.01 cid

Horsepower expresses the rate of doing mechanical work—pumping water, cutting grass, generating electrical power, and the like. It seems like a straightforward concept and has sold many an engine since James Watt coined the term in the 18th century. But the horses promised by the manu-facturer might not come when you summon them.

The power curve, i.e., the plot of horsepower against rpm, does not tell the complete story. Since 1987, Briggs engines have been rated according to the protocols established by SAE Code J-1349. The horsepower ratings are

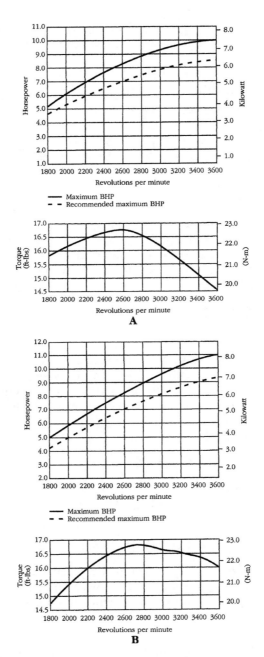

FIG. 1-7. *Horsepower and torque curves for two dimensionally similar engines: (A) the model 243400, a venerable-iron engine; (B) the more modern 252400.*

achieved during laboratory tests under standard environmental conditions. Briggs cautions that a prudent designer should not expect to have more than 80% of the rated maximum power available under real-world conditions of temperature, altitude, and engine tune.

Power depends on many factors, such as displacement, compression ratio, volumetric efficiency, and engine speed. The critical factor is engine speed, or rpm. Figure 1-7 shows power curves for two Briggs engines of almost identical displacement. One is a carryover from the days when engines were made of iron and develops 10 hp. The other represents a latter generation and produces a nominal 11 hp.

The 11-hp engine is, of course, more powerful than the antique, but that advantage is limited to the upper end of the rpm scale. At 2600 rpm, both produce slightly more than 8 hp. Drop the speed to 2000 rpm and the older engine begins to show the advantages of conservative valve timing. At 1800 rpm, the iron engine develops about ¼ hp more than its modern counterpart.

The shape of the power curve is at least as significant as rated horsepower. Hard-used engines need power at high rpm; industrial models that slog along at part throttle are better served by flat power curves.

Torque is the measure of instantaneous twisting force on an imaginary beam whose length is expressed in feet or meters. Horsepower is what one exerts when riding a bicycle at a steady speed on a level surface. Torque is the heave required to break a rusty bolt loose.

The ideal engine would have a flat torque curve throughout the rpm range, but that is difficult to arrange. Torque peaks at around three-quarters throttle. What happens to the curve on either side of the peak determines the flexibility of the engine and its speed sensitivity to load. Our perception of power, the way the engine refuses to stall under load, has more to do with torque than horsepower.

The iron engine builds torque like a locomotive up to a 2600-rpm peak then drops off sharply. The newer model doesn't do as well at low speeds, but holds its peak in an almost horizontal line to the rpm limit. This engine is obviously intended to operate at wide throttle angles.

The shape of the torque curve is influenced by valve diameter, port length and shape, and by the camshaft grind. A small-valved engine with restrictive porting and a conservative camshaft tends to make torque early and lose it at high rpm. An engine that flows better is happiest at higher speeds, where the flat torque curve also helps to build horsepower.

2

Troubleshooting

A gasoline engine needs ignition, fuel, and compression to run. But that's like saying a plant needs sunlight, nutrition, and water to grow. The terms need qualification to be useful for gardeners and mechanics.

Ignition must occur inside of the cylinder. A spark plug that fires outside of the engine may not function when the voltage requirement doubles under compression. Fuel is needed, but it must be present in the right amount and in the form of an explosive vapor. Too much fuel is as bad as none. And, finally, we need some compression to give combustion explosive force, but not so much as to make starting difficult.

The art of troubleshooting consists of making these often subtle qualifications. Some mechanics are very good at it; others merely throw parts at the problem. But troubleshooting is not difficult if you work from a baseline of known good parts and approach things in a systematic, step-by-step manner. Assume that the malfunction is simple and 99% of the time you will be right.

Work in a well-ventilated area, away from possible ignition sources. If you have to go into the fuel system, take the machine outdoors because gasoline will be spilled as fuel lines are disconnected and the carburetor bowl is removed. Gasoline vapor has a low flash point, very rapid flame propagation and, at around 19,000 Btu/lb, contains more energy than TNT. The PN 19051 ignition tester (shown in Fig. 2-1) or the PN 19368 are well worth the few dollars they cost. These tools confine the spark behind a window to prevent accidental ignition. Testing for spark in the old-fashioned way with a screwdriver can have unfortunate consequences.

Begin by eliminating as many variables as possible. Replace the spark plug with a new one of the correct heat range and reach. Most are gapped to 0.030 in., but there are exceptions as listed in Table 3-1 of the following

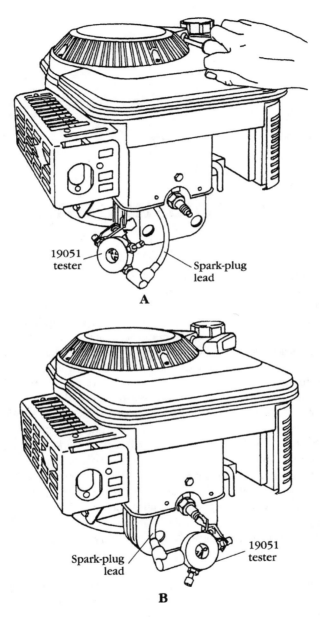

FIG. 2-1. *PN 19051 can be used to register the presence of ignition voltage during cranking (A) and to detect voltage interruptions in a running engine (B). In neither case does the tool say anything about spark-plug performance, which can only be determined by substitution of a known good plug. Also note that a spark gap in series with the plug (B) boosts voltage to help clear flooded engines.*

chapter. Once or twice in a working lifetime, a mechanic encounters a new spark plug that refuses to fire in small engines. Although the possibility of this happening is extremely remote, it is not a bad idea to test spark plugs in a running engine before using them as diagnostic tools.

Gasoline has a shelf life of less than 6 months, before toluene and other light hydrocarbons evaporate, leaving a sticky residue behind. Unless you are certain of fuel quality, drain the tank into a glass jar for examination. An acrid smell, water globules, or rust discoloration mean that you need to clean the tank and carburetor as outlined in Chapter 4. The EPA classes gasoline as a hazardous waste; disposal is best done by burning small quantities.

Clean or replace the air filter if it's dirty, as it almost surely will be. A restrictive filter element acts like a choke to enrich the mixture and will send you off on a merry chase making carburetor adjustments. Some heavy-duty engines have a paper filter in the fuel line between the tank and carburetor. Remove the filter and blow through it in the direction of fuel flow as indicated on the filter body. There should be little or no resistance. When in doubt, replace it.

While you are swapping out these parts, try to become acquainted with the engine in the same way that good doctors engage their patients in small talk. If the engine belongs to someone else, try to find out if it's been worked on since the last time it ran. If so, you can almost be sure that the mechanic made some sort of mistake.

Water is an enemy: large, flaky rust blisters on the muffler and corrosion on unpainted aluminum suggests that the machine was exposed to weather. Expect to find corrosion on electrical contacts and, quite possibly, water damage to the fuel system. Briggs gas caps are not rainproof. A loose blower housing on a rotary lawnmower—especially if the bolt holes in the sheet metal are elongated—points to a bent crankshaft.

Heavy accumulations of oil on the cooling fins of four-cycle engines usually means that the operator overfilled the crankcase, but it can also mean that the flywheel-side crankcase seal has failed. In any event, the shroud will have to be removed and the engine cleaned.

The crankcase oil level should register full on the dipstick or reach the top thread of the filler-plug boss. Black, carbonized oil that feels gritty when rubbed between the fingers is a sure sign of trouble. The engine might not be worth fixing. Sometimes the dipstick will have a stratum of clean oil at the top and black goo below. This means that the operator refilled the crankcase with fresh oil when the engine wouldn't start. But damage to internal parts has already occurred.

At this point, you should have a pretty good sense of the overall condition of the engine and have established a baseline with a functional spark plug, a supply of fresh fuel, and an unobstructed air filter. With luck, the problem will already be solved.

Failure to start

Electric starters can be had on engines as small as 5 hp. Make sure the battery is fully charged and use the starter sparingly, since the motors require a 15-minute cool down period between 1-minute engagements. Most mechanics prefer to use the rewind starter during diagnosis.

Ignition

When a cold engine refuses to start after three or four pulls of the starter cord, remove the spark plug, place the control lever in the "Run" position, and test ignition output with PN 19051. The tool has two electrodes, one with a 0.166-in. gap used for all modern engines and a 0.60-in. gap for testing the anemic Magna-Matic magneto used on 9, 14, 19, 23 cast-iron engines and on early production 191000 and 231000 series.

Connect the tester as shown in Fig. 2-1 with the larger gap in series with the ignition cable and an engine ground. Spin the flywheel vigorously and watch for spark. Magnetron pulse generators, which were introduced in 1982 and became universal by 1984, produce spindly spark best observed in the shade. The magnetos previously used should deliver a thick, blue spark of the type associated with automotive ignition systems.

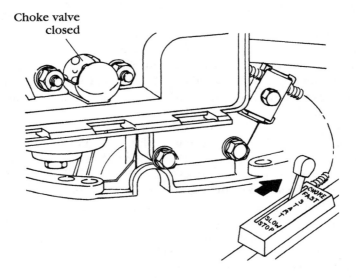

FIG. 2-2. *The choke, whether automatic or manual-remote as shown, must close fully for the engine to start.*

If there is no spark, look for:

- A sheared or badly distorted flywheel key. Replace the key with the correct factory part. See Chapter 3 for this and other ignition-system repairs.
- Magnetron or magneto failure.
- A shorted kill-switch wire, usually as the result of improper routing. Trace the wiring and repair.

Primer and choke

If the engine with a known good spark plug has ignition, the next most likely culprit is the cold-starting system. Fuel-tank mounted carburetors generally employ a primer pump to richen the mixture for starting. Remove the air cleaner and press the rubber bulb three or four times. The pump should inject a stream of gasoline into the carburetor bore. See Chapter 4 for repair information.

Other carburetors use a butterfly choke just aft of the air cleaner. The butterfly must close fully for cold starting (Fig. 2-2). The fuel systems chapter describes how to adjust automatic chokes found on Vacu- and Pulsa-Jet carburetors. Manual chokes engage by moving the control lever past the wide-open-throttle position. Adjusting the Bowden cable usually fixes the problem. A few Walbro carburetors employ a small helper spring that pulls the choke closed with an audible click. Verify that the spring is present and connected.

Fuel flooding

Any engine will flood if cranked long and hard enough, or if cranked when hot with the choke closed or with some other defect that prevents starting. As often happens, the mechanic floods the engine while trying to fix it.

To detect possible flooding, remove the spark plug, which should smell of gasoline and may be slightly damp. A dripping wet spark-plug tip is a sure sign that the combustion chamber and inlet pipe are flooded with raw fuel. An even more severe type of flooding occurs when the carburetor float mechanism sticks open. When that happens, gasoline pours out of the air horn or through the overflow tube on newer Walbro carburetors.

Clear the flooding by taking the machine outside and blowing out the cylinder with compressed air. Or simply wait an hour or so for the surplus fuel to evaporate. When attempting to start a flooded engine, use a dry spark plug and crank with the choke disengaged and throttle full open. We want to ingest as much air as possible. It's also helpful to connect the PN 19051 tester in series with the spark plug as shown back in Fig. 2-1B. The 0.166-in. gap boosts coil output by some 13,000 V.

A warm engine that vaporizes the fuel can make it difficult to detect flooding. The spark-plug tip will be dry, but the air-fuel mixture will be too gasoline-rich to ignite. If you suspect this is the case, allow 20 minutes or more for the vapor to disperse. Install a new spark plug and crank with the throttle wide open and the choke disengaged. If the engine were flooded, it should now be dry enough to start. At first there will be a few meager pops, then puffs of black smoke as the engine burns off the surplus fuel.

As in card games, there is a time to fold. Engines clear themselves of flooding, when one stops cranking and permits the collected fuel to evaporate.

Oil flooding

An oil-fouled spark plug means that crankcase oil has found its way into the combustion chamber. Any four-cycle engine that refuses to start after heroic cranking will pump crankcase oil past the rings. Rotary lawnmowers with their engines mounted head forward oil-flood when tilted up on their front wheels for servicing. The same effect can be produced by tilting the engine to the carburetor side. Oil runs into the carburetor through the crankcase vent line. The problem can be eliminated if you remove the filler plug or dipstick and use a hand pump to evacuate the crankcase.

Oil-flooding cures itself in an engine that is parked overnight. A quicker solution is to blow down the cylinder with compressed air. It can also be helpful to spray a small amount of Wynn's Carburetor Cleaner or an equivalent product into the carburetor air horn. Reinstall the air cleaner before engaging the starter.

Warning: Do not crank an engine without the air cleaner, which serves as a spark arrestor, in place. Carburetor cleaners increase the fire hazard. Ether-based starting fluid is even more volatile and is not recommended.

The starting drill requires three or four dry spark plugs. As mentioned earlier, a PN 19051 ignition tester connected in series with the spark plug boosts cranking voltage. Attempt to start the engine, replacing fouled plugs with fresh ones. Eventually, the engine should come to life in a cloud of blue smoke.

No fuel

If the spark plug remains dry after a half-dozen choke-on starting attempts, the problem is fuel starvation. The litmus test for fuel starvation is to remove the air cleaner, squirt a *small* amount of fuel into the carburetor bore, replace the air cleaner and crank. If the engine runs for a few seconds and quits, the problem is fuel delivery.

Warning: Using an oil can to squirt fuel into the engine is one of those procedures that factory manuals omit because of legal ramifications. People have been hurt doing this sort of thing. Many mechanics routinely prime carburetors with raw fuel, but these are often the same fellows who use their teeth for bottle openers. If you wish to follow their example, work outside, use a minimum amount of fuel, wipe up any spills, and be sure to reinstall the air cleaner before attempting to start the engine.

In rough order of frequency, fuel starvation comes about because:

- *Older Briggs float-type carburetors and most on-tank units.* These carburetors have adjustable jets that may have been tampered with. Make sure the adjustment screws are backed out two or three turns from lightly seated. See Chapter 4 for this and other fuel-system repair procedures.
- *Float-type carburetors.* Verify that fuel reaches the carburetor by cracking the input line. If not, work upstream to find the obstruction, which may be at the filter (when present) or at the screen integral with the tank fitting. If fuel reaches the carburetor, loosen the brass nut under the float bowl that secures the bowl to the carburetor body. Gasoline should dribble out around the nut. No fuel means that the float-controlled needle valve is not opening. Fuel in the bowl means a clogged main jet.
- *Tank-mounted carburetors.* Replace the carburetor diaphragm. All Vacu-Jet and some Pulsa-Jet pickup tubes incorporate a check valve that tends to stick closed.
- *Fuel-pump failure on engines so-equipped.* Test by cracking the input line to the pump. Fuel should be present. If not, the problem is upstream of the pump. If the pump sees fuel, then crack its output line and crank the engine with the ignition "Off." Fuel should be present. If not, change out the pump diaphragm.
- *Massive air leaks in the induction track.* caused by loose carburetor mounting screws. Inspect the flange gasket for tears and retighten the screws.
- *Insufficient compression.*
- *Failure of the intake or exhaust valve to open, isolating the cylinder.*

Loss of compression

Automatic compression releases make precise measurements of compression impossible. The most frequently encountered device, known as Easy-Spin, incorporates a "dead spot" on the cam that holds the intake valve a few thousands of an inch off its seat until late in the compression stroke. The loss of compression makes cranking easier, without costing much power. Other compression releases employ centrifugal flyweights to crack the exhaust valve open at low rpm or a starter-engaged yoke that has the same effect. Some large, single-cylinder engines use both types of compression release.

The factory-supplied cylinder bleed-down tool gives a pass-fail indication of compression and some insight into leak sources. But for our purposes it is enough merely to establish the existence of compression.

Short out the ignition with PN 19051 or the equivalent to prevent accidental starting. Crank the engine over. Resistance should build on the starter cord during the compression stroke, increase slightly at mid-stroke where piston velocity and friction are highest, and peak as the piston approaches top dead center. A dead feel to the cord means no compression. If you use an electric starter, listen to the sound, which should drop in pitch as compression is encountered. If the starter motor hums smoothly and the flywheel turns more rapidly than normal, you can be confident that the engine has lost compression.

Compression releases can be defeated by grounding the ignition and removing the blower housing. Spin the flywheel counterclockwise by hand, being careful not to cut yourself on the governor vane. The flywheel should rebound sharply against cylinder compression.

Zero or near-zero compression means:

- A blown-head gasket. See Chapter 7 for this and other mechanical repairs.
- A stuck-open valve or, in the worst case.
- A thrown rod.

Valve-related failure

One has to be unlucky to experience failure of the valve-actuating mechanism, but it can happen. Stale gasoline that coats the intake valve with sticky varnish or rust on the exhaust valve are the culprits. Valves normally stick closed against their seats.

What happens next depends upon the type of valve-actuation mechanism. Side-valve engines open their valves with tappets that ride directly on the camshaft. The first time the engine is cranked, the associated tappet forces the valve open. If the valve hangs, that is, if spring tension cannot pull it back down on its seat, the result is zero compression.

The story with overhead valves is different. Valves located above the cylinder head are opened by small-diameter pushrods that, like soda straws, bend easily. During cranking, the pushrod buckles, leaving the valve stuck tightly against its seat. Depending upon the valve involved, the air-fuel mixture cannot enter the cylinder or, once in, cannot escape. Compression, as measured by our rough indicators, appears normal.

This condition can be quickly detected by removing the valve cover. A bent pushrod will have slipped out from under its rocker arm.

Engine runs 3 or 4 minutes and quits

Connect PN 19051 in series with the spark plug and start the engine. Watch the arcing in the window to determine if ignition failure initiates shutdown. When this happens, the flywheel coasts to a stop without generating spark.

Another possibility is a kind of delayed-action fuel starvation, which can occur with carburetors that draw from a diaphragm-supplied reservoir on the top of the tank. If this is this case, the spark-plug tip will be bone white. Replace the carburetor diaphragm. A clogged fuel-cap vent has the same effect.

Loss of power

Lack of power can be a bit tricky to diagnose. The problem may be with whatever equipment the engine drives. Make sure drive belts have some slack and turn driven equipment through by hand to detect possible binds.

A stretched or distorted governor spring can limit engine rpm and power. Replace the spring with the exact factory part, of which there are dozens, to prevent possible overspeeding.

Warning: Use the correct PN throttle spring and do not attempt to bend governor links or make adjustments to the mechanism. Governor adjustments are one of the few jobs that are best left to authorized service centers. In no case should the engine be run at wide-open throttle and without load.

Another possibility is a high-speed ignition miss. Connect our old standby PN 19051 in series with the spark plug and crack the throttle open to detect possible misfiring. If spark activity is constant, the next step is to examine the spark-plug tip color. A lean mixture stains the tip bone-white and may produce pop-backs and flat spots during acceleration. Most carburetors have one and sometimes two adjustment screws. Richen the mixture as described in the fuel systems chapter. If adjustments do not help and the carburetor-mounting screws are tight, dismantle and chemically clean the carburetor.

An overly rich mixture leaves fluffy carbon deposits on the plug and, when extreme, smokes the exhaust. Rich engines have a soft exhaust note. Assuming that the air filter is clean and that the choke opens fully, adjust the carburetor as necessary to lean out the mixture.

The correct mixture leaves brownish black deposits, the color of coffee with a dash of cream, on the spark-plug tip.

Failure to idle

Small air-cooled engines do not like to idle. Some run at constant factory-set speed; those with variable throttles idle, if that is the correct term, at between 1700 and 1750 rpm. Adjust the carburetor as described in Chapter 4 and verify that carburetor mounting screws are secure. If the problem persists, clean the carburetor with special attention to the low-speed circuit.

Failure to shutdown

All models employ a single-strand ground wire running from the ignition module to a kill switch mounted near the carburetor or as part of the flywheel brake. A misrouted ground wire can pull loose from the switch or break, making it impossible to short out the ignition.

Remote throttles that operate by means of a Bowen cable frequently get out of adjustment. Loosen the locking screw and reposition the armored cable so that the remote control lever fully closes choke at one extreme of travel and shorts the ignition at the other.

Kickback

Kickback feels as if someone were trying to jerk the starter cord out of your hand. Severe kickback can break flywheel ring-gear teeth on electric-start models. The most likely culprit is a sheared or distorted flywheel key that acts to advance the ignition. Some magneto-fired models and 23, 24, and 32000 Magnetron-equipped engines have adjustable timing that can be improperly set. Another possibility is failure of the automatic compression release to unseat its associated valve, either because of failure of the mechanism itself or because there is too much valve clearance.

Kickback often occurs with inexpensive discount-house rotary lawn-mowers. These machines depend upon the blade to provide the flywheel mass necessary for smooth starts. If the blade is missing or loose, the engine kicks back. Most of these engines run rich and develop heavy carbon deposits in the cylinder. Removing these deposits restores the compression ratio to normal and seems to help starting.

Starter binds

Rewind starters bind and can fail to retract when misaligned with the flywheel cup. Loosen the hold-down bolts and reposition the blower housing as necessary. Check tooth engagement on electric-start models as described in Chapter 5. Binding can also be caused by loose motor hold-down bolts, wear on armature bushings or a bent motor shaft.

Exhaust smoke

Acrid, black smoke means an overly rich air-fuel mixture. Verify that the choke opens fully, the air filter is clean, and that the carburetor is adjusted correctly.

Blue smoke results from oil in the combustion chamber. Check for too much oil in the sump or oil that has been diluted with gasoline. Worn engines may smoke with multigrade oil and be tolerable with 30W-grade. The crankcase should be under a slight vacuum, produced by the check valve in the breather. Chapter 7 describes how to test for crankcase vacuum. A loose dipstick cap or failure of the o-ring at the connection between the dipstick-tube and engine block will destroy the vacuum and permit oil to enter the chamber.

The classic cause of blue smoke is piston-ring wear, although overhead value engines will smoke upon starting and, sometimes, during acceleration when the intake-valve seal has failed.

Excessive vibration

It's the nature of single-cylinder engines to vibrate. Excessive vibration of the kind that loosens bolts and produces wheel rumble on lawnmowers, and generally makes things unpleasant is almost always the result of a bent crankshaft. But first check that motor hold-down bolts are tight and that there are no fatigue cracks radiating from the bolt bosses.

If we're dealing with a rotary lawnmower (the great crankshaft bender), empty the tank to prevent fuel spillage, remove the spark plug, and tilt the machine up on its front wheels. As mentioned earlier, engines mounted head-forward will oil-flood unless the crankcase is first pumped out.

Warning: Do not work on the underside of a rotary mower without removing the spark plug or grounding the spark-plug lead with PN 19051. Moving the blade, even inadvertently, can start the engine.

Mark a point on the underside of the deck adjacent to a blade tip. Rotate the flywheel 180° and see the other blade tip registers with the mark previously made. If not, the blade and/or the crankshaft are bent. To determine which is at fault, remove the blade and adapter, and place the blade on a flat surface. Bends or twists should be obvious. Next have a helper pull the starter cord while you focus on the bolt hole in the end of the crankshaft. If the hole appears to wobble as the crankshaft spins, the crank is bent and should be replaced.

A more precise measurement can be made by sanding the rust off the crankshaft stub and measuring runout with a dial indicator. The amount of bend is one-half of the total indicated runout. In other words, if the indicator shows a total displacement of 0.020 in., the amount of bend is 0.010 in., or ten times the permissible 0.001 in.

3

Ignition systems

Ignition-system failures are common, although the solid-state spark generator has improved things immensely by eliminating contact points. Major
offenders are the spark plug, which should be replaced at the first sign of
trouble and the flywheel key.

Spark plugs

Spark plug recommendations vary, depending upon the source and the date
the information was compiled. As can be best made out, all OHV engines
leave the factory with resistor-type Champion RC 12YC plugs. The more
durable platinum-tipped version of this plug, standard on I/C engines and
available from Briggs dealers as PN 5066, can be substituted. Side-valve
engines now use RJ19LM plugs, except for models 19A400 through 19G400.
These 9- and 10-hp engines require RC14YC. While heat ranges may vary
slightly, plugs from other manufacturers can be substituted:

Champion RC12YC = Autolite 3924 Bosch FR8DCX

Denso Q16PR-U

Champion RC14YC = Autolite 3926

Bosch FR8DC

NGK FR5

Champion RJ 19LM = NGK BR2LM

Spark plugs are gapped at 0.030 in. except for Intec models 11060, 11160,
120600, 121600, and 122600 that have a .020-in. gap. Why this variation
exists is something that only Briggs & Stratton engineers know for sure.

Reducing the spark gap advances the ignition a tad and should enable the engine to start at lower cranking speeds.

Most people find the wedge-type gauge, like the one shown in Fig. 3-1, more convenient to use than the wire type, although the latter is a better choice for used plugs with concave side electrodes.

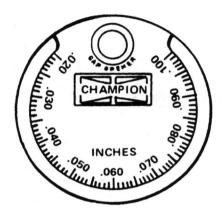

FIG. 3-1. *Champion spark-plug gauge, available at auto-parts stores.*

The interface between the spark plug and the cylinder head acts as a heat sink. Remove all traces of oil and grease from this area. Some mechanics apply a squirt of silicone—a moderately effective lubricant—to the spark-plug threads prior to assembly. Motor oil, which carbonizes under the heat, is not recommended. Run the new plug in three turns by hand and torque to 180 in./lb for all models, except side-valve engines built before 1981. The aluminum heads on these older engines require 240 in./lb of spark-plug torque and iron heads need 330 in./lb.

Battered or carboned-over threads (caused by using an oiled spark plug or one with insufficient reach) can be chased with an M14 × 1.25 metric tap. Stripped threads call for the services of an automotive machinist with access to the necessary HeliCoil tools.

Flywheels

The flywheel must be lifted to check the condition of the key and to service magneto contact points and condensers. Disconnect and ground the spark-plug lead to prevent accidental starting. If the engine is equipped with a 12-v electric starter, disconnect the positive (red) battery terminal. Remove the cooling shroud together with the rewind starter. Depending upon the engine model, this can entail removing the gas tank, oil-filler tube, starter motor, and other hardware. A nut, washer, and tapered fit secure the flywheel to the crankshaft. The key is a sacrificial item used as a marker for ignition timing.

The starter clutch acts as the hold-down nut on the traditional Briggs rewind starter that remains in production on the "Classic" line of engines. The clutch requires a special wrench PN 19114 or, preferably, PN 19161 (Fig. 3-2). The latter can be used with a torque wrench during assembly. All other engines secure the flywheel with a hex nut. Crankshaft threads have been right handed since the early 1950s.

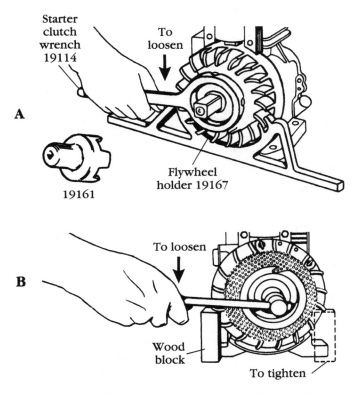

FIG. 3-2. *A starter clutch (A) or hex-nut (B) secures the flywheel. If you have to buy a special tool for the starter clutch, the torque-wrench compatible socket PN 19161 is the better choice than PN 19114. Note that Briggs cast-iron engines, fitted with notched pulleys in lieu of rewind starters, had left-hand threads. When in doubt, trace the lay of the threads that extend past the flywheel nut.*

If you don't have an impact wrench, you must make provision to hold the crankshaft stationary while the flywheel nut is loosened. In any event, the flywheel must be secured to apply the proper amount of torque upon assembly.

Caution: Do not attempt to block the flywheel with a screwdriver. This almost invariably results in broken impellor blades and an imbalanced flywheel.

There are various ways to safely prevent a crankshaft from turning. A two-by-four inserted between the blade and deck works for direct-drive rotary mowers. The factory tool, shown in Fig. 3-2A, fits 6.75-in. aluminum wheels.

Iron wheels are strong enough to tolerate wedging with wood block (Fig. 3-2B). A large strap wrench, like the one shown in Fig. 3-3, fits all wheel sizes. Old-time mechanics, who had little to work with except their native intelligence, would run a length of rope down through the spark-plug port.

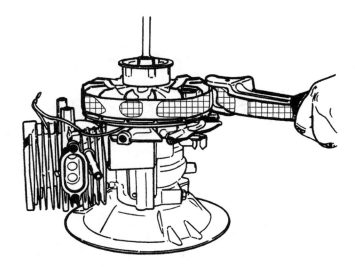

FIG. 3-3. *At strap wrench, used here on a Tecumseh engine, has almost universal application.*

The next task is to separate the flywheel from the crankshaft, which are bound together by the taper fit. Aluminum flywheels have two unthreaded bosses on the hub to give purchase to the self-tapping bolts supplied with Briggs pullers (Table 3-1). Tightening the bolts lifts the flywheel. But strip a

Table 3-1
Factory flywheel pullers

PN	Model range
19069	side-valve 60000 through 120000
19165	side-valve: 140000, 170000, 190000, 1975 build date, and earlier 250000
19203	side-valve: 220000, 230000, 240000, post-1975 build date 250000 through 320000 OHV: 97700, 99700, 110400, 110600, 111400, 111600, 113400, 120400, 120600, 121600, 122600, 123400, 123600

hub thread and you are in a world of trouble. My personal preference is to cut the threads with a 5/16 in. × 18 in. tap and use a Kohler-type tool that applies torque to the crankshaft (Fig. 3-4).

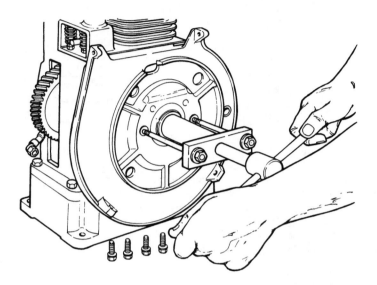

FIG. 3-4. *A Kohler flywheel puller fits most small engines and applies torque to the crankshaft stub, rather than to vulnerable flywheel-hub threads.*

Caution: Do not shock the flywheel loose with a crankshaft knocker or brass bar. This procedure pits ball-bearing races, scrambles flywheel magnets, and poses the risk of crankshaft breakage. Briggs could, in theory, void the warranty on engines abused in this fashion.

Iron flywheels come free by gently tapping the back side as the crankshaft is rotated. Do not strike areas adjacent to the magnets.

Inspect the flywheel key for damage. The key locates flywheel magnets relative to the ignition coil. This relationship determines the timing for engines with Magnetron pulse generators and synchronizes flux buildup with point opening for magnetos. A few thousandths of an inch of key distortion translates into a major error at the flywheel rim. A sheared key usually prevents any spark from being generated.

The aluminum key also functions as a shear pin to protect more expensive components. Never use a steel key on Briggs engines. The new key should fit snugly into the keyways on both the flywheel and crankshaft. There is no method of restoring worn keyways other than to replace the affected parts. However, one can cobble up a repair as described in Chapter 7.

Examine the flywheel for damage with particular attention to the hub area (Fig. 3-5).

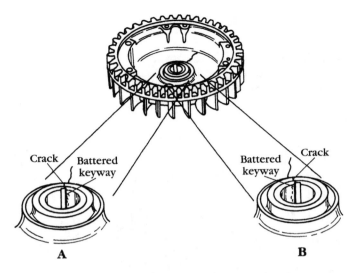

FIG. 3-5. *A crack or wallow on the leading edge of the keyway (in the direction of rotation) suggests that the flywheel was loose and overrun by the crankshaft (A). Damage to the trailing edge of the keyway means that the flywheel attempted to overtake the crankshaft (B). This can only occur if the crankshaft suddenly stopped, as when a rotary lawnmower blade strikes something solid. In either case, the flywheel key and possibly the crankshaft must be replaced.*

Table 3-2
Flywheel torque and air gap, single-cylinder engines

Engine type/ model name	Cubic-inch displacement (1st one or two digits of model number)	Armature air gap— two legged (in.)	Armature air gap—three legged (in.)	Flywheel-nut torque (lb/ft)
Side-valve, aluminum block	6, 8, 9, 10 (except as noted below), 12	0.008	0.014	50–55
	100200, 100900, 13	0.012	0.014	60–65
	17, 19, 22, 25	0.012	0.014	70–75
	28	0.012		100–105
Side-valve, cast-iron block	23, 24, 30, 32	0.012	0.023	140–145
Europa OHV	9	0.008–0.009		60–65
Intec OHV	11, 12	0.010–0.012		60–65
Intec OHV	20	0.008–0.009		110–115
Intec OHV	31	0.010–0.012		95–100
PowerBuilt *	28	0.012–0.014		95–100

* No data on other PowerBuilt models.

Warning: Always replace a cracked or otherwise damaged flywheel. Once a crack starts, it continues to grow until the critical length is reached. At that point, the wheel explodes.

Clean all traces of grease and oxidation from the tapers, and carefully file off any burrs that might be present. If a Bellville washer is used under the nut or starter clutch, replace it with dished, or concave, side toward the flywheel. Torque to specification as shown in Table 3-2. It goes without saying that an impact wrench should not be used for this operation. These wrenches develop enough torque to crack the flywheel.

Magnetos

A DIY mechanic occasionally encounters a magneto-fired engine and needs some idea of how to proceed. Check the flywheel key as described earlier. Next check the condition of the points, which on smaller engines live under the flywheel. Oil in the point housing will defeat ignition and usually enters via a failed crankshaft seal or wear on the plunger that activates the points. Wear on the plunger bore can be corrected if you find a Briggs dealer with the proper reamer. The seal can be replaced without disassembling the engine as described in Chapter 7.

Replace the point set and condenser before doing anything else. These parts can still be found, but have become expensive. Figure 3-6 shows the more common point set with the "hot" contact as part of the condenser.

Note how this little jewel goes together with a depressor tool (packaged with the parts) used on the condenser (Fig. 3-7) and the braided ground wire looped over the post that supports the grounded contact (Fig. 3-8).

Once you have installed the point set, make a preliminary adjustment so that the points open a few thousandths and close when the crankshaft keyway aligns with the plunger. Adjustment procedures are described in the captions for Figs. 3-9 and 3-10. All Briggs and, for that matter, Tecumseh, Clinton, Power Products, West Bend, and Pincor magnetos have the same 0.020-in. gap specification, although some seem to run better with 0.018-in.

Oil from the feeler gauge or even fingerprints on the contacts can deny ignition. Burnish the contacts with a business card. Remove the spark plug, ground the ignition with PN 19051 or the equivalent tool, and spin the flywheel. The magneto should produce a thick, blue spark. If not, burnish the points and retest.

No spark after repeated point massaging, means that it is time to do some serious inspection. The stop switch should be insulated from engine ground when the control lever is the "Run" position. Some coils have an external ground wire connected by a screw to the laminations. Make sure the mating surfaces are clean and bright. Check for continuity between the spark-plug

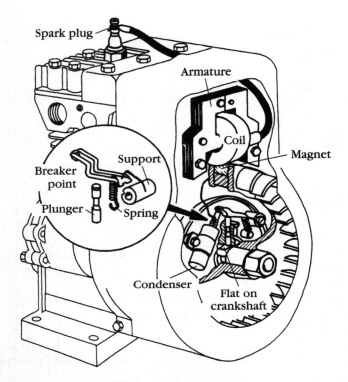

FIG. 3-6. *Millions of Briggs & Stratton under-flywheel magnetos continue to give service. The most popular variant employs the point-and-condenser set shown.*

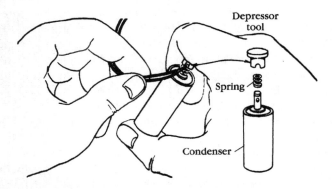

FIG. 3-7. *A depressor tool, packed with replacement point sets, should be in every mechanic's tool box.*

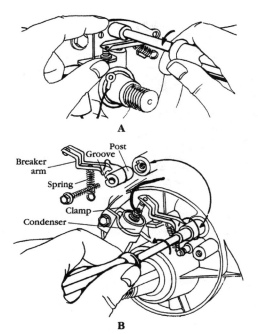

FIG. 3-8. *The earlier under-fly-wheel point set employs a remotely mounted condenser wired to the moveable arm (A). In the later version the condenser doubles as the grounded contact (B). The braided ground wire should be routed as shown, over the top of the post. The open eyelet of the spring attaches to the moveable point arm.*

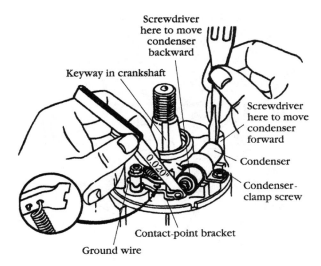

FIG. 3-9. *Turn the crankshaft to bring the keyway in line with the point plunger, snug down the condenser-clamp screw and set the gap at 0.020 in. Use a screwdriver to move the condenser as necessary. Tighten the condenser-clamp screw and recheck the gap, which almost invariably will have changed. Repeat the operation, this time compensating for the effect of the clamp screw.* Briggs & Stratton Corp.

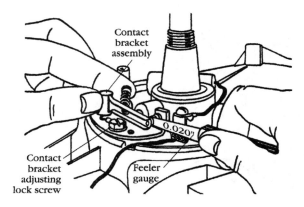

FIG. 3-10. *Briggs two-piece point sets have a screwdriver slot to facilitate adjustment. Otherwise, the drill is the same as described in the caption for Fig. 3-9.*

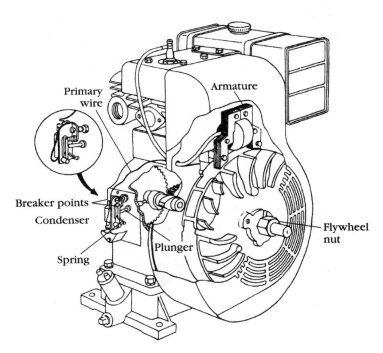

FIG. 3-11. *An external point set and ignition coil were used on horizontal-shaft engines in the 19–32 cid displacement range. Both this and the Magna-Matic were troubled by oil intrusion into the point chamber. Coat the point and condenser hold-down and cover screws with nonhardening sealant and, if you can find the part, install an oil seal over the outboard end of the point plunger.* Briggs & Stratton Corp.

terminal and the coil connection. Verify that the air gap between the coil and flywheel is within specification (see the section "Armature air gap."). As a last resort, replace the coil.

Briggs also supplied magnetos with camshaft-activated points (Fig. 3-11 illustrates one version). The most impressive was the Magna-Matic, used on large horizontal-shaft engines and identified by a horseshoe magnet, rotor, and coil all mounted under the flywheel (Fig. 3-12). The next drawing, Fig. 3-13 shows the point set, awesome in its complexity. Elongated stator mounts permit the timing to be advanced and retarded, but timing marks should be present. Figure 3-14 illustrates the rotor adjustment for late production Magna-Matics.

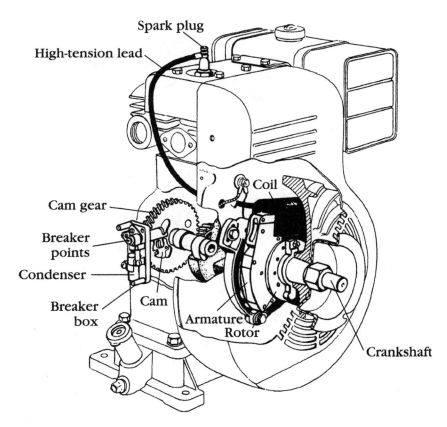

FIG. 3-12. *The Magna-Matic has an under-flywheel coil, a rotor, and external points.* Briggs & Stratton Corp.

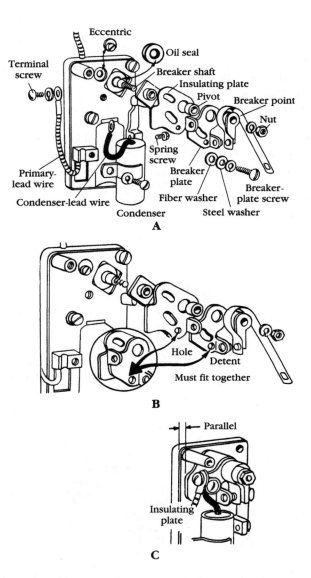

FIG. 3-13. *The Magna-Matic point set cannot be beaten for shear complexity (A) Install the breaker plate with the detent on the bottom of the stationary point bracket in the matching hole in the insulating plate (B) Failure to do this will warp the assembly and misalign the contacts. Turn the eccentric screw to bring the left edge of the insulating plate parallel to the edge of the box (C) The breaker is spring-loaded to advance the ignition. Turn the shaft against spring tension to its stop. While holding the shaft in that position, slip the moveable point arm over the shaft, install the lock-washer and tighten the nut. Rotate the flywheel to open the points and, using the eccentric screw, set the gap at 0.020 in. Apply sealant to breaker-plate screw threads and tighten. Recheck the point gap.*

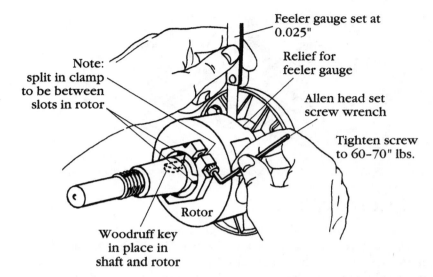

FIG. 3-14. *On late production units, the rotor stands off 0.025 in from the main bearing to allow the crankshaft to float without damage to the rotor. Earlier units were snugged together without clearance. The clamp also serves as a keeper to complete the magnetic circuit between poles. The clamp and rotor should remain together during storage. Units are stamped with three timing marks identified by engine displacement.*

Armature air gap

Table 3-2 lists armature air gaps for all engines under discussion. The reference in the table to two- and three-legged armatures may need clarification. Most magnetos have their coils wound on two-legged, U-shaped armatures. Others used three-legged armatures shaped like the letter E.

The narrower the gap, the better the spark. But some clearance must be allowed for main-bearing wear, flywheel eccentricity, and thermal expansion. Skid marks on the rim of the wheel mean the gap was too narrow. Severe interference makes itself known initially as misfires, and then, as the coil melts, as loss of ignition.

To set the gap, rotate the flywheel until the magnets are adjacent to the armature. Insert the appropriately sized feeler gauge or shim stock between the magnets and armature, loosen the armature hold-down screws, and retighten (Fig. 3-15). Remove the gauge and check the clearance at several points on the flywheel rim.

Many mechanics do not bother with gap specifications, which to some extent are arbitrary. They use a playing card instead.

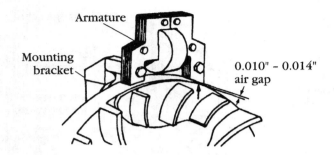

FIG. 3-15. *Shim stock can be used to set the air gap as shown.* Briggs & Stratton Corp.

Magnetron

If the coil is outboard of the flywheel and has only one small wire running from it to the kill switch, the device is a Magnetron (Fig. 3-16). Introduced in 1982 and made universal within a year or two, the unit consists of an ignition coil and a transistorized switch triggered by a flywheel magnet. Early versions had a replaceable trigger module; later versions integrate the module with the coil.

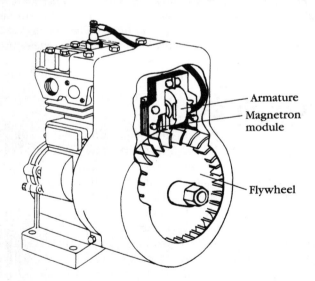

FIG. 3-16. *The Magnetron module has one small-diameter primary wire running from it to the engine stop switch.* Briggs & Stratton Corp.

The Magnetron uses a conventional ignition coil consisting of primary and secondary windings on a laminated iron armature and encapsulated in epoxy, similar in appearance and function to the coil shown in Fig. 3-17. In

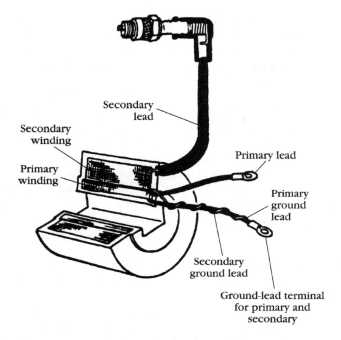

FIG. 3-17. *The Magnetron ignition coil is fundamentally no different than coils used with other systems.* Tecumseh Products Co.

addition, the unit incorporates a third, or trigger, winding and two transistors. These paired transistors are somewhat confusingly referred to in the singular as a "Darlington transistor" (Fig. 3-18).

As a flywheel magnet approaches the armature, magnetic lines of force permeate the laminations to induce voltage. The Darlington transistor, excited by the voltage in its trigger coil, becomes conductive and completes

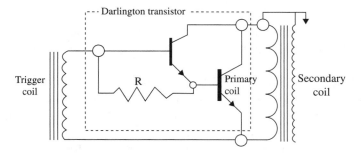

FIG. 3-18. *The Magnetron trigger circuit is similar to those used in many automotive applications.*

the primary circuit to ground. Further movement of the flywheel saturates the primary with magnetic flux. Fairly high currents, on the order of three amps, flow through the primary. These currents generate a powerful magnetic field that saturates the secondary windings. In other words, the primary winding acts like an electromagnet.

As the flywheel continues to turn, magnetic polarity reverses. Trigger voltage changes polarity and the Darlington transistor no longer conducts. The resistor shown in the schematic builds voltage to speed the shut down process. Denied ground, current no longer flows in the primary. The primary windings cease to act as an electromagnet. The collapse of the magnetic field, which takes place at near light speed, induces high voltage in the secondary. This voltage finds ground by arcing across the spark-plug gap.

Output is normally in the 6–10,000 volt range, although the Magnetron can deliver much higher voltages. The ignition transformer has 74 turns on the primary and 4400 on the secondary to give a theoretical voltage multiplication of 59.5.

The Magnetron incorporates an ignition advance. At low speeds, the flywheel magnet must be within close proximity to the trigger coil in order to induce the 1.2 v needed to activate the Darlington transistor. At higher speeds, the magnetic threshold is lowered and ignition occurs earlier.

Magnetron coils are stamped with the year, month, and day of manufacture. If the unit came with the engine, its date will be a month or so earlier than the engine build date.

Magnetron service

Test system output with PN 19051 or 19368 as described in the Troubleshooting chapter. Magnetrons need a vigorous pull on the starter cord to generate visible spark. Discrete trigger modules for early models can be replaced for about $20. This was a good feature, since Darlington transistors often fail because of overheating. Resistance readings on the coil secondary (i.e., between the spark-plug cable terminal and engine ground) should be between 3000 and 5000 ohms. Newer Magnetrons are nonrepairable and retail on the net for about $40.

Install the unit with the spark-plug cable up. Set the air gap, connect the kill-switch lead and torque hold-down screws to 25 lb/in.

Magneto-to-Magnetron conversion

This section describes how to replace points and condenser with a Magnetron trigger module. These modules are compatible with most aluminum-block single-cylinder engines with two-legged coil armatures.

Certain older engines, many of them iron-block collector's items, remained in production during the early 1980s when they were modified by the factory for solid-state ignition. Owners of magneto-fired versions of these engines often want to follow the factory's example, but the conversion can be costly.

Magnetron-compatible flywheels, with the necessary third magnet, can cost $250. Check the prices at the Briggs web site before you embark upon such a project.

The PN 394970 kit includes a trigger module and miscellaneous bits of hardware. Follow this procedure to install the module on compatible magneto-fired engines.

1. Remove the spark plug and disconnect the battery on electric-start models.
2. Lift the flywheel to access primary-circuit and kill-switch wiring. You may snip the wires at the dust cover and leave the points and condenser undisturbed. (Fig. 3-19). Or, to make a neat job of it, remove the points and condenser, and seal the point-plunger hole in the crankshaft with PN 231143 or the equivalent. Failure to plug the hole results in a major oil leak.

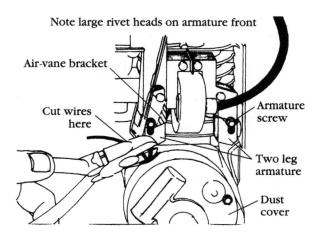

FIG. 3-19. *It's not necessary to remove the points and condenser when retrofitting a Magnetron module. Note that the large rivet heads define the outboard, or flywheel side of the armature.* Briggs & Stratton Corp.

3. Remove the two cap screws that secure the ignition coil and, on some applications, the governor vane bracket, to the engine. You might have to cut part of the bracket away for module clearance (Fig. 3-20).
4. Stand the coil armature upright on its legs with the engine side facing you. The heads of the rivets that hold the armature together should not be visible. As the coil is now oriented, the trigger module mounts between it and the right leg of the armature.
5. Slip the module into position without forcing it. Clip the plastic hook over the armature shoulder.

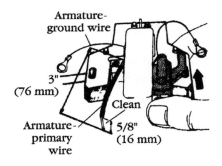

Armature-ground wire

3"
(76 mm)

Clean

Armature-primary wire

5/8"
(16 mm)

FIG. 3-20. *The module installs on the right-hand leg as viewed from the underside of the armature.* Briggs & Stratton Corp.

6. Two and sometimes three primary wires originally ran from the point assembly, one to the coil and the others to external engine shutdowns. One end of another primary wire is connected to the trigger module with the other end free. These primary wires can be identified by the presence of insulation. An uninsulated wire connects the coil windings to the armature by means of a screw. Another bare wire with a terminal on the end comes out of trigger module. These bare wires are grounds.

7. Peel back the insulation 3/4 in. from the ends of the wires that ran from the point set. Scrape away the varnish.

8. Route these wires under the ignition coil and up to connect with the module. If the wires are too short, splice in short pieces of insulated primary wire supplied with the kit. Solder the joints with 60–40 rosin-core solder and insulate with heat-shrink tubing. Keep splices as short as possible.

   Caution: Do not use acid-core solder or crimp-on connectors on these or other connections.

   Once the primary wires are long enough to reach the trigger module, twist them together two turns and solder. The connection should look like the letter "Y."

9. Strip off 1/2 in. of insulation and varnish off the end of primary wire coming from the module.

10. The installation kit includes a small coil spring and a T-shaped retainer with a hook on the end. Slip the spring over the long arm of the T, so that it shoulders on the cross bar (Fig. 3-21). Place the retainer—hook first—into the recess on the side of the module.

11. Compress the retainer spring with a Briggs condenser or a 3/16-in. drill bit, as shown in Fig. 3-22. Insert the end of primary wiring, which you soldered together in Step 8, together with the primary wire from the module, under the head of the retainer. This connection ties the primary ignition circuit to the trigger module and to whatever engine stop switches may be present. Release spring tension and, using

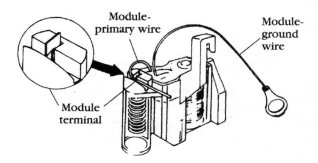

FIG. 3-21. *A clip-and-spring assembly connects primary wires to the trigger module.* Briggs & Stratton Corp.

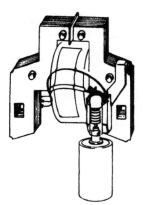

FIG. 3-22. *Use a Briggs PN 294628 condenser or a drill bit to compress the spring on the trigger-module connection.*

long-nosed pliers, rotate the retainer to engage its hook into the slot on the module. Trim the wires about 3/16 in. out from the retainer.

12. Twist the module-ground and the coil-ground wires together. Solder the connection. Both of these uninsulated wires have terminals on their ends. Snip off the terminal on the shorter one.

13. Secure the insulated wires to the underside of the coil with Permatex No. 2 or equivalent to prevent vibration and wire breakage.

14. Mount the coil on the engine. Use the mounting screw on the right side of the armature as the ground connection (Fig. 3-23). Do not attach the armature/module ground wire to the air-vane bracket.

15. Set the air gap as described previously and torque the mounting screws to 25 lb/in.

16. Using the key provided in the kit, install the flywheel on the crank-shaft. Connect a PN 19051 or PN 19368 ignition tester between the spark-plug cable terminal and engine ground. Spin the flywheel vigor-ously by hand.

Warning: Contact with the air vane can result in a nasty cut.

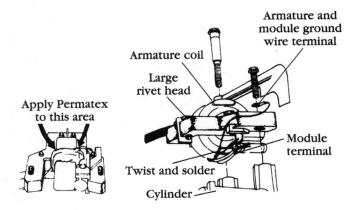

FIG. 3-23. *Secure the primary wires to the underside of the coil with silicone or equivalent sealant. Module and ignition-coil ground wires are twisted together and soldered. Use the armature hold-down screw on the leg opposite to the air-vane bracket for the ground.* Briggs & Stratton Corp.

17. If no spark, check the wiring. Verify that all insulated wires are connected and secured to the module by the spring clip, and that the two bare wires are connected and grounded by an armature hold-down screw. Soldered joints should be bright and smooth.
18. Torque the flywheel to specification and complete the assembly. Test the engine.

Replacing an existing switch module

The replacement procedure follows the previous outline, except that breaker points and condenser are, of course, not part of the picture. Be careful not to overheat the plastic module retainer when unsoldering the wires (Fig. 3-24).

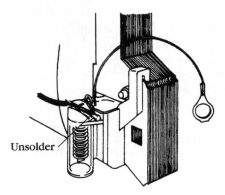

FIG. 3-24. *Exert care when unsoldering wires from the module.* Briggs & Stratton Corp.

Shutdown switches

Briggs uses a variety of engine shutdowns, but all work by grounding the primary side of the ignition circuit. For some applications, a remotely mounted lever controls choke engagement, engine rpm, and shutdown through a Bowden cable. Others omit the choke function, which is either automatic or in the form of a primer pump on the carburetor. Rotary lawnmowers combine engine shutdown with a flywheel brake that engages when the deadman's lever is released. The kill switch is usually integrated with the brake mechanism, but some applications have the switch on the carburetor. Regardless of location, all stop switches for engines covered in this book employ a copper tang as the grounding element and secure the wire from the coil with spring clip.

Failure to shutdown

This malfunction is usually caused by an improperly adjusted Bowen cable (Fig. 3-25). Place the throttle lever or deadman's switch in the "Stop" position and verify that the part labeled "Control lever" or the equivalent part on the brake mechanism makes contact with the copper grounding tang. If the switch is loose, replace it. Models that use the same cable to engage the choke require some judicious adjustment to shut the choke completely at one extreme of travel and to ground ignition at the other. Also check that the ground wire is connected to the copper tang. The single-strand ground wires Briggs uses have been known to break inside of the insulation. If you suspect this is the problem, place the control lever in the "Run" position and detach the wire at the ignition-coil spade connection. Test continuity between the free end of the wire and the copper tang with an ohmmeter. The meter should read something approaching zero resistance. If the meter shows infinite resistance or if resistance increases as the wire is flexed, the wire is broken and should be replaced.

Loss of ignition

When there is no spark and the usual suspects—the spark plug, flywheel key and, when fitted, the contact points—have been exonerated, check for a shorted kill-switch circuit. With the control lever in the "Run" position, disconnect the ground wire at the coil and test for continuity between the wire and a good paint- and grease-free engine ground. Zero resistance means a short in the wire or grounding tang.

Safety interlocks

Interlocks vary, since the original equipment manufacturer is responsible for their design. Some include logic modules to defeat tampering. From a mechanics point of view, interlocks fall into two categories: those that are

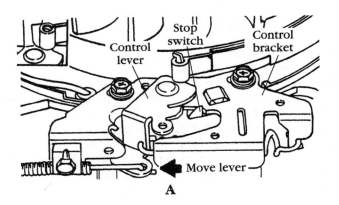

A

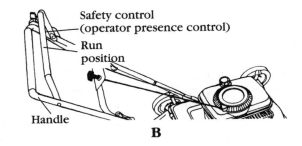

B

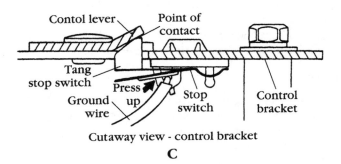

Cutaway view - control bracket

C

FIG. 3-25. *One approach Briggs took to the "compliance engine" was to integrate the handlebar safety control with the existing stop switch mounted on the carburetor. To test, place the safety-control lever in the "Run" position (A) remove hands (B) and verify that the control lever makes contact with the stop-switch tang (C). The same verification can be made on models with the stop switch integrated into the flywheel brake mechanism by removing the cooling shroud.*

normally open (NO) and those that are normally closed (NC). NO interlocks shunt primary-side ignition current to ground when closed. NC contacts open to interrupt the primary current. In either event, the engine stops.

The surest way to test interlocks is with an ohmmeter. Isolate the suspect component from the circuit and measure its resistance in the normal and tripped modes. Figure 3-26 illustrates another technique used for simple circuits without anti-tampering devices. If the engine refuses to start with interlocks in the circuit and runs with a jumper connected as shown, the source of the problem is obvious.

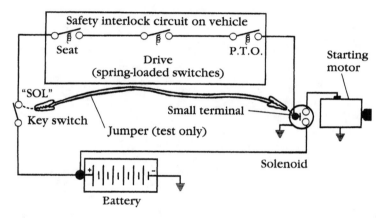

FIG. 3-26. *Typical safety interlocks.*

4

The fuel system

The fuel system consists of the carburetor, air cleaner assembly, fuel line, tank, and miscellaneous fittings. Optional elements include a fuel pump (which might be integral with the carburetor), shutoff valve, and filter.

How carburetors work

We think of an engine as a source of power. From the fuel system's point of view, the engine is a vacuum pump. The partial vacuum created by the piston during the intake stroke sets up a pressure differential across the carburetor. Air and fuel, impelled by atmospheric pressure, move through the instrument to equalize pressures.

Venturi & high-speed circuit

If you look through a carburetor, you'll see that the bore has an hourglass shape, with the necked-down portion located just upstream of the throttle plate. This area is known as the venturi. As much air leaves the carburetor as enters. Consequently, air velocity through the venturi must be greater than through the straight sections of the bore on either side of it. The increase in velocity is purchased at the expense of pressure.

Fuel, under atmospheric pressure, moves from the carburetor reservoir through the main jet and into the nozzle, which opens to the low-pressure, high-velocity zone created by the venturi. The jet can be fixed, as shown in Fig. 4-1A, or adjustable. In either case, the size of the jet orifice determines the strength of the mixture by regulating how much fuel passes into the venturi. The main air jet, more commonly known as the main air bleed,

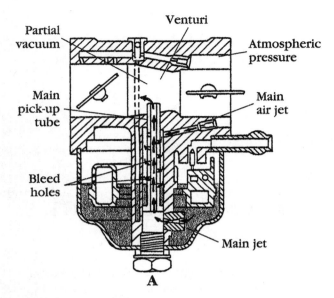

FIG. 4-1. *At wide throttle angles, fuel enters the venturi through the main pick-up tube, better known as the* high-speed nozzle. *(A) An air bleed ("main air jet") emulsifies the fuel, breaking it into droplets prior to discharge.* Briggs & Stratton Corp.

emulsifies the fuel before discharge, primarily to prevent siphoning. Collectively, these parts make up the high-speed circuit—high speed because the venturi works only so long as the throttle is open. Closing the throttle blocks air flow through the venturi and shuts down the fuel circuit.

Throttle & low-speed circuit

The throttle blade, or butterfly, controls engine speed by regulating the amount of fuel and air leaving the carburetor. It functions like a gate valve, opening for the engine to develop full power and almost completely blocking the bore at idle. The restriction generates a low-pressure zone downstream of the throttle blade, exactly as if it were a venturi. Fuel enters through a series of ports drilled in the carburetor bore. The port nearest the engine—known as the primary idle port—functions when the throttle blade is against its stop (Fig. 4-1B). As the throttle cracks open, one or more secondary ports are uncovered to ease the transition between idle and main venturi startup (Fig. 4-1C). The low-speed circuit also includes an air bleed.

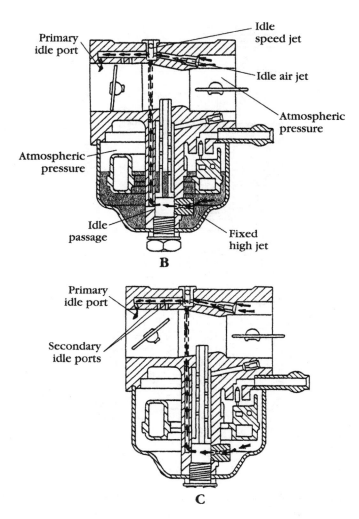

FIG. 4-1 (Cont.). *When the throttle blade is closed, the low-speed circuit discharges into the primary idle port. As the blade opens, it uncovers secondary idle ports (C). Wider throttle angles generate flow through the main jet.* Briggs & Stratton Corp.

Fuel inlet

Carburetors always include a mechanism for regulating the internal fuel level, independent of delivery pressure. Most Briggs type carburetors employ a float-actuated inlet valve, known as a *needle and seat* (Fig. 4-1D). When the reservoir is full, the float forces the needle against its seat, which cuts off

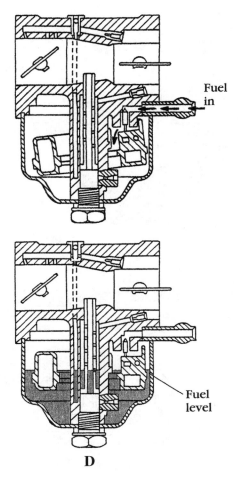

FIG. 4-1 (Cont.). *A float-operated inlet valve, or the needle and seat, maintains a constant fuel level in the instrument (D).* Briggs & Stratton Corp.

fuel delivery. As gasoline is consumed, the float drops, which releases the needle and opens the valve.

Suction-lift carburetors draw from the tank through a pickup tube in a manner analogous to a flit gun (Fig. 4-2). A check valve prevents fuel from draining out of the tube during starting.

Cold start

With one or two exceptions, Briggs carburetors employ conventional choke valves to enrich the mixture during cold starts. The Walbro carburetor shown in Fig. 4-1A uses a pivoted choke disk. The Vacu-Jet in Fig. 4-2 has

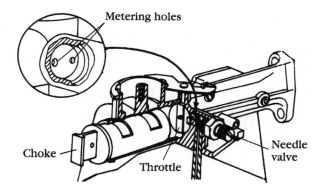

FIG. 4-2. *Vacu-Jets and their Pulsa-Jet cousins draw through a pickup tube that extends into the fuel tank. An internal check ball prevents fuel from running back out of the tube. Note the plug-type choke and single-adjustment needle that controls both low- and high-speed mixture strength.* Briggs & Stratton Corp.

an old-fashioned plug choke. In either case, closing the choke seals off the carburetor bore. The engine, in effect, pulls on a blind pipe. All jets flow in response to the low pressure.

Figure 4-3 illustrates the Briggs automatic choke found on Vacu- and Pulsa-Jet carburetors. A spring-loaded diaphragm holds the choke closed

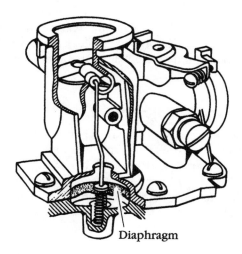

FIG. 4-3. *Some Vacu- and Pulsa-Jet carburetors are fitted with a vacuum-operated automatic choke. The choke butterfly should close when the engine is not running, flutter shut under sudden load and acceleration, and open at steady speed. Major causes of failure are a bent air cleaner stud, leaking diaphragm, and dirt in the butterfly pivots. The diaphragm chamber must be airtight.* Briggs & Stratton Corp.

during cranking. Upon starting, a manifold vacuum, acting on the underside of the diaphragm, overrides the spring to open the choke. The choke also operates as an enrichment valve should the engine falter under load; the loss of the manifold vacuum allows the choke to close.

Some automatic choke mechanisms are fitted with a bi-metallic helper spring. A tube connected to the breather assembly conducts warm air over the spring, which causes it to uncoil and open the choke independently of the vacuum signal acting on the choke diaphragm.

Pulsa-Prime nylon-bodied carburetors combine a fuel pickup tube with a primer pump. No choke is used. Pressing the primer bulb evacuates air from the pickup tube, which causes the level of fuel in it to rise.

External adjustments

Figure 4-4A illustrates a Flo-Jet carburetor with adjustable main and low-speed jets. Backing out the needle-tipped adjustment screws opens the jet orifices and enriches the mixture (Fig. 4-4B). Tightening the screws makes the mixture leaner by restricting the fuel flow through the jet. The idle-speed adjusting screw bears against the throttle stop to regulate idle rpm.

All Briggs carburetors have an idle speed adjusting screw. Most of the newer types dispense with one or both of the mixture-adjustment screws, as shown in Fig. 4-1A. Fixed-main and/ or low-speed jets should require no attention unless the calibration is upset by a change in altitude. An engine set up by the factory for sea-level operation will run rich in the rarefied air at high altitudes. Briggs can supply the correct jetting. A field fix that works on some B & S-supplied (and other) Walbro carburetors is to remove the air-bleed jet.

Initial mixture screw adjustment

As a rule of thumb, engines should start when the mixture-control screws are backed out $1^1/_4$ to $1^1/_2$ turns from lightly seated.

Caution: Use your thumb and index finger to seat the screws. Do not force the issue with a screwdriver. Adjustment screws that have been damaged by overtightening must be replaced if the engine is to run properly (Fig. 4-5).

Procedure

With a clean air filter in place, fully open choke, and fresh fuel in the tank, run the engine under moderate throttle for about five minutes to reach operating temperature. The tank should be about half full on engines equipped

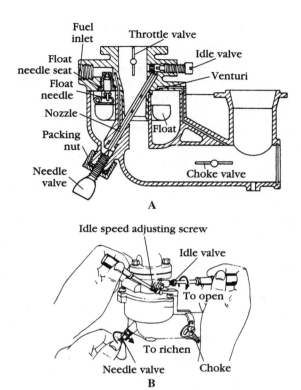

Fuel inlet
Throttle valve
Idle valve
Float needle seat
Venturi
Float needle
Nozzle
Packing nut
Float
Needle valve
Choke valve

A

Idle speed adjusting screw
Idle valve
To open
To richen
Needle valve
Choke

B

FIG. 4-4. *Two-piece Flo-Jet in sectional view (A). The part that Briggs calls a "needle valve" is better known as the main, or high-speed, mixture-control screw. The "idle valve" is the idle, or low-speed, mixture-control screw. The locations of these two mixture-control screws varies with carburetor type but the idle-mixture screw is always the one closer to the engine. Because these screws control fuel delivery, backing them out of their jets enriches the mixture (B). Occasionally, you might encounter a foreign or vintage American carburetor that employs adjustable air jets, distinguished by the rounded needle tips of the adjustment screws. Backing out an air screw leans the mixture.* Briggs & Stratton Corp.

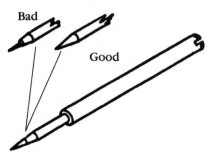

Bad

Good

FIG. 4-5. *Bent or grooved adjustment needles must be replaced.* Briggs & Stratton Corp.

with Vacu-Jet carburetors to minimize the effects of fuel level on mixture strength. This requirement does not apply to other carburetors.

1. Run the engine at about three-quarters speed.
2. Back out the main mixture-control screw in small increments—no more than an eighth of a turn at a time. Pause after each adjustment for the effect to be felt. Stop when engine rpm drops and, using the screwdriver slot as reference, note the position of the screw at the rich limit.
3. Tighten the screw in increments as before. Stop when engine speed falters at the onset of lean roll, which represents the leanest mixture that supports combustion.
4. Open the adjustment screw to the midpoint between the onset of lean roll and the rich limit.
5. Close the throttle and adjust the idle mixture for the fastest idle. You need at least 1700 rpm.
6. Snap the throttle butterfly open with your finger. If the engine hesitates, back out the high-speed adjustment screw a sixteenth of a turn or so and repeat the experiment. Enriching the high-speed mixture usually calls for a slightly leaner idle mixture.

The adjustment always imposes some compromise between idle quality and high-speed responsiveness. This is especially true for Vacu- and Pulsa-Jet carburetors that regulate both mixtures with a single adjustment screw. Always err on the side of richness, and do not consider any carburetor adjustment final until proven under load.

Troubleshooting

Make the checks described in chapter 2 before assuming something has gone amiss with the fuel system. Of course, sludge in the tank or raw gasoline dribbling from the air horn are powerful arguments for immediate action.

No fuel delivery

The engine appears to develop compression and the carburetor is not obviously loose on its mountings. The spark-plug tip remains dry after prolonged cranking. When mixture screws are present, backing them out has no effect. An injection of carburetor cleaner through the spark-plug port brings the engine back to life, but only for a few seconds.

Tank-mounted, Vacu-Jet carburetors suffer total failure when the check ball in the fuel pipe sticks in the closed position, almost always as a result of stale gasoline. Its cousin, the Pulsa-Jet, quits when its fuel pump diaphragm stretches or ruptures. The high- and low-speed jets (actually discharge ports) in these carburetors do not often clog and hardly ever do so simultaneously.

Float-type carburetors are susceptible to blockages between the tank and carburetor-inlet fitting. Possibilities include a clogged tank screen, fuel-tank cutoff valve, or filter on engines so equipped. External and crossover Flo-Jet fuel pumps might also fail because of an internal malfunction or a vacuum leak. (Older engines sometimes used mechanical pumps, which are susceptible to diaphragm and check-valve failure.) Check fuel delivery by cracking the line at the carburetor inlet.

If no fuel is present, work backwards, connection by connection, to the tank.

Warning: Opening fuel lines is always hazardous and especially so if the engine must be cranked to activate a fuel pump. Make these determinations outdoors with the ignition switched off and the spark-plug lead solidly grounded.

If fuel appears at the carburetor inlet, it should also be inside the instrument. Remove the fuel bowl, which is secured by a central nut on the underside. The bowl should be full. If not, the problem is a stuck inlet needle, hung float, or clogged inlet screen (on units with this feature). Failure to transfer fuel out of the bowl suggests a clogged main jet, fuel-delivery nozzle, or loss of manifold vacuum.

Engine runs lean at full throttle

This fault will appear as loss of power, possible backfire as the throttle plate is suddenly opened, and a dead-white or bleached-brown spark-plug tip. The engine seems to run better when choked. Backing out the high-speed adjustment screw (when present) has no effect.

Begin by looking for air leaks downstream of the throttle plate. Focus on the carburetor mounting flange and cylinder head gasket. Lean running in a two-cycle engine is the classic symptom of crankshaft seal failure, but don't settle on this rather grim diagnosis until other possibilities have been eliminated.

Replace the optional fuel filter and open the line to verify that copious amounts of fuel are available at the carburetor inlet fitting. Note the preceding warning about ignition sparks and spilled gasoline. Finally, look for a stoppage in the high-speed circuit, which will usually be at the point of discharge.

Engine runs rich

A blackened spark plug, acrid, smoky exhaust, and loss of power suggest an overly rich mixture. It is assumed that turning the mixture adjustment screws (when present) has no effect and that the air filter has been cleaned or, if made of paper, replaced. The choke butterfly opens fully, as verified by visual inspection with the air cleaner removed.

In my experience, persistently rich mixtures are a problem almost entirely confined to float-type carburetors. Replace the needle and seat and set the float level to specification. If the difficulty persists, check for clogged air bleeds.

Briggs & Stratton Walbro carburetors use a replaceable high-speed air jet that might have been removed in an ill-advised attempt to richen the mixture.

Note: Fixed-jet carburetors require recalibration at high altitudes. Contact your dealer for recommended fuel and air jet sizes.

Carburetor floods

Long bouts of cranking with the choke closed will flood any carburetor, wet the bore, and spill fuel out of the air horn. Two-piece Flo-Jets flood quite easily.

Note: Carburetor flooding, characterized by fuel puddling in the bore, must be distinguished from external leaks. Fuel will cascade past a worn or twisted float-bowl gasket. Tank-mounted Vacu- and Pulsa-Jet carburetors might weep fuel at the tank interface because of cracks in the tank or a bad gasket. See the following "Removal and installation" section for additional information.

Failure of the inlet needle and seat or of the float mechanism produces spontaneous flooding in Flo-Jet and B & S Walbro carburetors. Needle-and-seat failures are usually attributable to wear, although dirt in the fuel supply can produce the same effect. Dirt-induced flooding might spontaneously cure itself, only to reappear as another particle becomes trapped between the needle and seat. Clean the fuel system and replace the needle and seat.

Float failures are usually of the obvious mechanical sort and correctable by cleaning.

Engine refuses to idle

Engines that operate under a constant-load regime might not have provision for idle. Once the engine starts, the governor raises engine speed to a pre-set rpm. This discussion applies to engines that left the factory with an idle capability and now refuse to exercise it.

Check for air leaks at the carburetor mounting flange, cylinder-head gasket, and at the throttle-shaft pivots. The latter source might not be significant but will allow abrasives to enter the engine. Other possibilities include:

- *Idle rpm set too high.* All Briggs & Stratton carburetors have an adjustable throttle stop in the form of a spring-loaded screw. Use a tachometer to adjust to the factory or equipment manufacturer's specification.
  Caution: Air-cooled, splash-lubricated engines do not idle in the automotive sense of the word. Speeds of 1700 rpm and more are the norm.
- *Maladjusted throttle cable.* Loosen the cable anchor and reposition the Bowden cable as necessary.
- *Binding throttle shaft or linkage.* New throttle shafts sometimes bind because of paint accumulations. Grass or other debris might limit the freedom of movement of the throttle-return mechanism.

- *Governor failure.* With the engine running and the throttle lever set on idle, gently try to close the throttle. Do not force the issue. If light finger pressure does not move the throttle against its stop, a possible governor malfunction is indicated for engines intended to run at variable speeds.
- *Clogged low-speed circuit.* Clean the carburetor.

Removal & installation

The carburetor bolts to the engine block or makes a slip-fit connection, sealed by an O-ring, with the fuel-inlet pipe. The fuel supply must be shut off on gravity-fed systems, either with the valve provided or by inserting a plug into the carburetor end of the flexible fuel hose.

Warning: Some gasoline will be spilled. Work outside in an area remote from possible ignition sources.

The governor mechanism must be disengaged from the throttle arm without doing violence to the associated springs and wire links. Some springs have open-ended loops and can be easily disengaged with long-nosed pliers. Others incorporate double-ended loops that come off and go on in a manner reminiscent of the "twisted-nail" puzzle.

Note the lay of the spring and, if there is any possibility of confusion, mark the attachment holes.

Wire links remain connected until the carburetor is detached from the engine. While holding the carburetor in one hand, twist and rotate it out of engagement with the links, being careful not to bend the wires in the process.

Four, and on one model five, screws secure Vacu- and Pulsa-Jet carburetors to the fuel tank. Automatic-choke models used on 920000, 940000, 110900, and 111900 engines have one of the screws hidden under the choke butterfly. Inspect the tank interface for cracks (in which case, the tank must be replaced) and for low spots that might cause fuel or vacuum leaks (Fig. 4-6). This problem is serious enough for the factory to supply a Pulsa-Jet tank repair kit (PN 391413).

Assembly is the reverse of disassembly. Always use new gaskets mated against clean flange surfaces.

Warning: Briggs & Stratton engines built before the early 1970s (the factory spokesperson contacted could not be more specific about the date) used asbestos gaskets. Soak the area with oil and remove gasket material with a single-edged razor blade and dispose of the shards safely. Do not use a wire wheel or any other method that might create dust.

Make up the links first, then secure the carburetor to the engine. Connect the springs last.

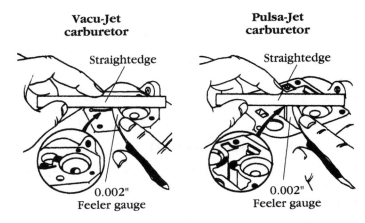

FIG. 4-6. *Vacu-Jet and Pulsa-Jet tank flanges are crucial. Hairline cracks or low spots leak fuel. A low spot in the shaded area shown, denies vacuum to the Pulsa-Jet fuel pump.* Briggs & Stratton Corp.

Automatic choke   Most late-production Vacu- and Pulsa-Jet carburetors are equipped with a type of automatic choke unique to Briggs & Stratton. The choke valve is normally closed by a spring and pulled open by a manifold vacuum acting on a large diaphragm.

The choke butterfly should snap closed when the engine stops and open as soon as it starts. If should also flutter in response to changes in the manifold vacuum under severe loads. In this function, the automatic choke acts as an enrichment device.

Spring length is critical:

Application	Spring free length
Vacu-Jet	$^{15}/_{16}$–1 in.
Pulsa-Jet	$1\,^{1}/_{8}$–$1\,^{7}/_{32}$ in.
Engine models 11090 & 11190	$1\,^{15}/_{16}$–$1\,^{3}/_{8}$ in.

To assemble:

1. If a new diaphragm is being installed, attach the spring as shown in Fig. 4-7.
2. Invert the carburetor and guide the spring/link assembly into its recess (Fig. 4-8).
3. Turn the assembly over and start the mounting screws, turning them just far enough for purchase.
4. Depress the choke butterfly with one finger and attach the actuating link (Fig. 4-9). This action preloads the diaphragm spring.

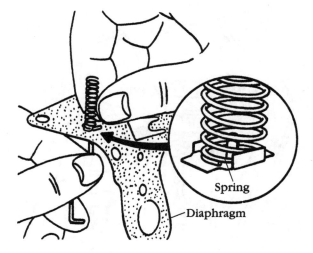

FIG. 4-7. *Attach the choke link to the diaphragm as shown.* Briggs & Stratton Corp.

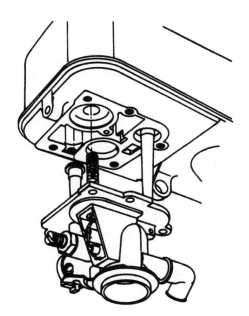

FIG. 4-8. *With the parts inverted, mate the carburetor to the tank. Turn the assembly over and lightly start the hold-down screws. The parts must be free to move relative to each other.* Briggs & Stratton Corp.

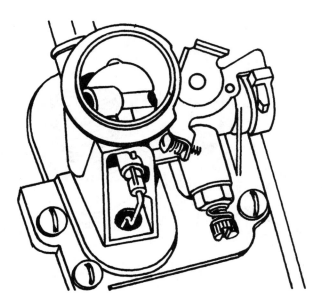

FIG. 4-9. *Complete the assembly while holding the choke closed.* Briggs & Stratton Corp.

5. Tighten the mounting screws, running them down incrementally in an X-pattern. Spring preload should lightly tip the choke over center, toward the closed position. Install the link cover and gasket.

If the choke remains closed after the engine starts, the problem can usually be traced to insufficient spring preload. Too much spring preload will hold the choke open, regardless of the strength of the vacuum signal. Fuel or oil in the diaphragm chamber has the same effect. Other possibilities include carbon/varnish deposits on the butterfly pivots or a bent air-cleaner stud. If the stud is bent, replace it, rather than attempting a repair that might compromise the air cleaner-to-carburetor seal.

Pulsa-Jet tank diaphragm Pulsa-Jet carburetors include a fuel pump, which is actuated by a vacuum diaphragm. Side-mounted diaphragms are discussed in an upcoming section titled "Vacu-Jet & Pulsa-Jet." The tank-mounted diaphragm (Fig. 4-10), used on one version of this carburetor, concerns us here.

Figure 4-6 illustrates the places where vacuum leaks typically develop at the tank interface. Figure 4-10 shows the sequence of assembly. Note that the spring rests collar down on the top of the diaphragm and not in the tank cavity.

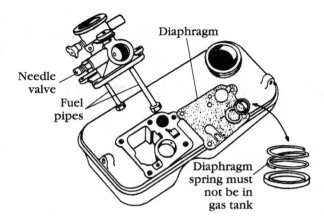

Needle valve

Fuel pipes

Diaphragm

Diaphragm spring must not be in gas tank

FIG. 4-10. *Pulsa-Jet diaphragms install on the tank followed by the wear collar and Spring.* Briggs & Stratton Corp.

Repair & cleaning

Carburetors do not wear out in the accepted sense of the term. Most failures are associated with "soft" parts, such as inlet needles and seats, diaphragms, and gaskets. After long service, the throttle shaft bearings might develop enough play to justify replacement, when such repairs are possible.

Light varnish deposits respond to lacquer thinner and compressed air. Heavier deposits can be removed with Gunk Carburetor Cleaner, which as these products go, appears to be fairly benign. But no chemical cleaner, however aggressive, can undo the effects of water-induced corrosion. White powdery deposits and leached castings mean that the carburetor should be discarded. Sometimes you can get a water-damaged carburetor to work, but it will never be quite right.

Strip off nonmetallic parts, including gaskets, O-rings, elastomer inlet seats, and diaphragms. Leave nylon butterflies in place, since a short exposure to Gunk does not seem to hurt. Nylon-bodied carburetors can also be cleaned, provided immersion is limited to five minutes or so.

As you clean and repair, don't disturb the following items:

- The throttle and choke butterfly valves.
- The Vacu- and Pulsa-Jet pickup tubes (unless for replacement).
- Any part that resists ordinary methods of disassembly. Threaded, as opposed to permanently installed pressed- fitted parts, have provisions for screwdriver or wrench purchase. But forced removal might do more damage than carburetor cleaner can correct.
- Expansion plugs (unless loose or leaking). Briggs & Stratton carburetor overhaul kits include expansion plugs. If you opt to replace these

items, install them with a flat- nosed punch (a piece of wood will do) sized to the plug's outside diameter (OD). Seal by running a bead of fingernail polish over the joint after installation.

Carburetor service by model

Depending on displacement, market, and vintage, Briggs & Stratton engines employ any of five basic carburetor types. Each type includes subvariants and most have undergone running production changes, which are keyed to the engine build date.

Two-piece Flo-Jet

Small, medium, and large two-piece Flo-Jets are the only up-draft carburetors in the Briggs line. Figure 4-4A shows the parts layout common to all three models.

Needle & seat   Extract the elastomer seat by threading a self-tapping screw into the fuel orifice (Fig. 4-11). Press in a replacement seat—PN 230996 for gravity feed, PN 231019 for applications with a fuel pump—flush with the casting. Viton-tipped needles can be reused, if not obviously worn.

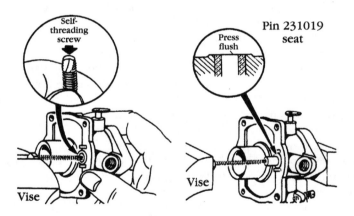

FIG. 4-11. *Briggs & Stratton uses elastomer inlet seats, which are extracted with a self-tapping screw and pressed into place using the original seat as a buffer. In the case of the one-piece Flo-Jet, the seat should be flush with the raised casting lip.*

Float setting   When inverted and assembled without a bowl-cover gasket, the float should rest level with the casting (Fig. 4-12). Adjust float height by bending the tang with long-nosed pliers. Figure 4-13 illustrates the proper orientation of the needle clip for this and other carburetors that employ the device.

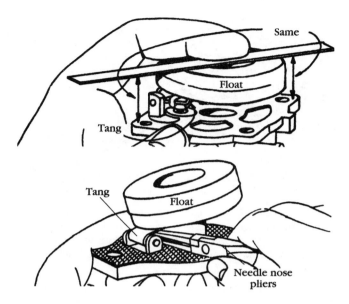

FIG. 4-12. *The float should be level with the (gasketless) casting for Flo-Jet and all other Briggs carburetors. Adjust by bending the tang without applying force to the needle.*

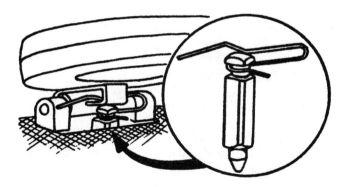

FIG. 4-13. *Install needle spring as shown for all applications.* Briggs & Stratton Corp.

Caution: Do not make the float height adjustment by pressing the float against the needle.

Casting distortion Overtightening the hold-down screws distorts the bowl-cover casting. Assemble without a gasket, and check with a 0.002-in. feeler gauge (Fig. 4-14). If the blade enters, remove the cover and straighten the "ears" with light hammer taps.

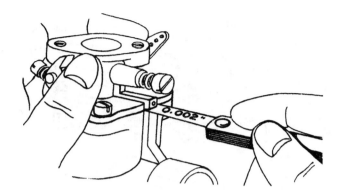

FIG. 4-14. *The two-piece Flo-Jet bowl-cover casting (or "throttle body") is a fragile casting, easily warped by over tightening. Allowable distortion is <0.002 in., as measured with a 0.5-in.-wide feeler gauge. If the gap is excessive, turn the casting over and, using a small hammer, tap the corners back into alignment.* Briggs & Stratton Corp.

Throttle shaft/bearing replacement   Follow this procedure:

1. Using a $1/8$-in. punch, drive out the roll pin that secures the throttle lever to its shaft.
2. Scribe marks on the throttle blade and carburetor body as assembly references.
3. Remove the two small screws that secure the throttle butterfly to the shaft.
4. Remove the butterfly and shaft.
5. Extract the shaft bushings with a $1/4$-in. tap.
6. Press in new bushings to original depth.
7. Install a new throttle shaft and throttle butterfly with scribe marks indexed. Coat the butterfly screw threads with Loc-tite. Start the screws, but do not tighten.
8. Close the butterfly to center it in the carburetor bore and tighten the screws. Verify that the butterfly swings through its full arc without interference. Complete the assembly.

One-piece Flo-Jet

Briggs manufactures one-piece Flo-Jets in two sizes. The small version carries its main jet in the float bowl (Fig. 4-15A). The large model employs a remotely located main jet, supplied through a removable nozzle (Fig. 4-15B).

Needle & seat   Service as described for the two-piece Flo-Jet (Fig. 4-11). Note that the replacement seat must be dead flush with the casting.

Float setting   Adjust as described for the two-piece Flo-Jet. When inverted, the float should rest level with the carburetor body (Fig. 4-12).

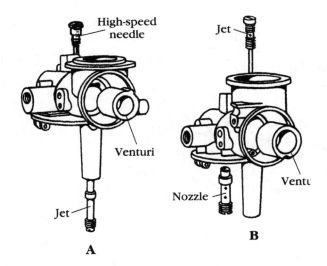

FIG. 4-15. *Small (A) and large (B) one-piece Flo-Jets with float bowls removed. The large model has its high-speed mixture screw under the float bowl.* Briggs & Stratton Corp.

Crossover Flo-Jet

Figure 4-16 is a sectional view of the crossover Flo-Jet, as used on horizontal-crankshaft Model 253400 and 255400 engines. This carburetor includes a vacuum-operated fuel pump, illustrated in the next drawing.

Needle and seat renewal and float adjustment procedures are as described for the two-piece Flo-Jet. Figure 4-17 illustrates the assembly sequence for

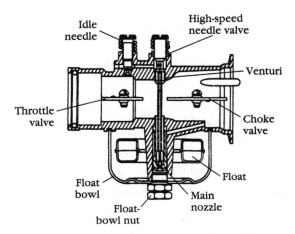

FIG. 4-16. *Crossover one-piece Flo-Jet adjusts from the top.* Briggs & Stratton Corp.

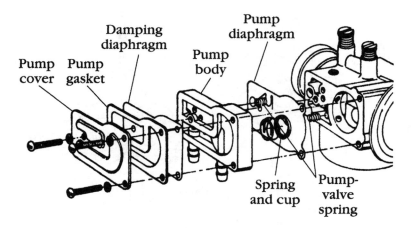

FIG. 4-17. *Briggs supplies replacement diaphragms singly or as part of the crossover Flo-Jet rebuild kit. Assemble dry, without sealant.*

the double-diaphragm fuel pump, normally serviced with a rebuild kit that contains springs and "soft" parts.

Briggs & Stratton Walbro

The factory appears to be slowly phasing out the Flo-Jet series in favor of highly modified Walbro carburetors. Engines in the 9-to 13-cubic-in. range use variants of the small series Walbro, recognized by its angular appearance and removable air bleed jet, mounted just aft of the air cleaner. The removable main jet nozzle has been omitted, together with replaceable throttle-shaft bushings and other niceties associated with earlier designs. Most small Walbros have a fixed-main jet and an adjustable low-speed jet, but there are exceptions. Some carburetors, as shown in Fig. 4-18, employ fixed jets for both circuits. Carburetors used on 120000 engines are completely adjustable.

The large B & S Walbro, fitted to 19-, 25-, and 28-CID vertical shaft engines, employs an external air bleed jet, which is located next to the idle mixture screw, and a removable main nozzle, which is accessed from the float bowl. Fixed-main jets are the norm.

Aside from the nozzle detail, service procedures are similar for the small and large versions of the carburetor.

Inlet seat   Fish out the elastomer seat with a hooked wire and install the replacement to cavity depth (Fig. 4-19) using a flat punch sized to seat OD.

Float   Float pins might include an antirotation feature in the form of flats milled on one end. Drive out the pin from the unmarked end and assemble with

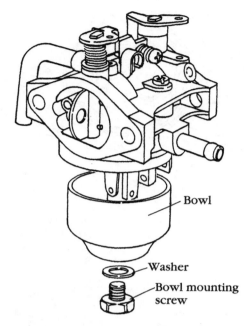

FIG. 4-18. *Because of emission regulations, nonadjustable carburetors are becoming the norm. The L.-shaped tube on the left vents the float bowl.* Briggs & Stratton Corp.

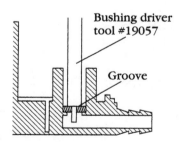

FIG. 4-19. *Press the fuel-inlet seat home with the grooved side of the seat down, toward the incoming fuel stream. A flat punch, sized to seat OD, can be used in lieu of the factory tool shown.*

the pin flats aligned to corresponding flats on the pivot bearing (Fig. 4-20). A spring clip ties the needle to the float (Fig. 4-21). Float height is fixed by needle and seat geometry and should not be tampered with in the field.

Vacu-Jet & Pulsa-Jet

Figure 4-2 illustrates the Vacu-Jet mechanism, which is distinguished by a single pickup tube and tandem discharge ports controlled by flow through a single jet. Figure 4-22 shows the three basic forms of this carburetor.

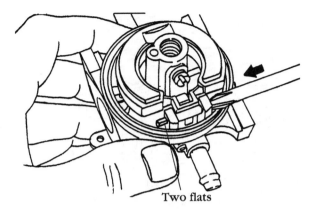

Two flats

FIG. 4-20. *Flats are provided on some float pins, apparently to prevent hanger wear. Extract from the flattened end and install with flats properly indexed.* Briggs & Stratton Corp.

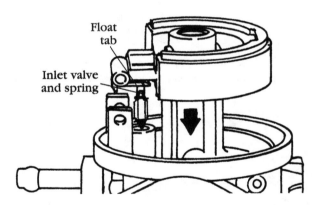

Float tab

Inlet valve and spring

FIG. 4-21. *Install the float with the needle assembled to its clip. Do not tamper with the float tang.* Briggs & Stratton Corp.

The Pulsa-Jet derives from the Vacu-Jet and in its various permutations uses many of the same parts. The distinction between the two is that the Pulsa-Jet feeds from a reservoir in the top of the fuel tank, which it replenishes with a vacuum-powered fuel pump (Fig. 4-23). Pulsa-Jets have two pickup tubes. The longer one transfers fuel from the tank to the reservoir; the shorter tube draws from the reservoir into the carburetor. This arrangement isolates the carburetor from changes in the level of fuel in the tank. Vacu-Jets lean out as the tank depletes.

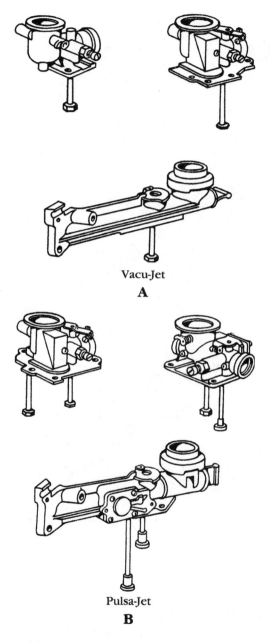

Vacu-Jet

A

Pulsa-Jet

B

FIG. 4-22. *Vacu-Jet (A) and Pulsa-Jet (B) variations.* Briggs & Stratton Corp.

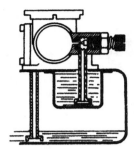

FIG. 4-23. *Three pulls of the starter cord should pump enough fuel into the Pulsa-jet reservoir to start the engine. Once it starts, the reservoir remains full to a level defined by a spill port. Thus, the internal fuel level of the carburetor is independent of the level of fuel in the tank. According to Briggs & Stratton, a Pulsa-jet-equipped engine develops as much power as one supplied by a more expensive float-type carburetor.*

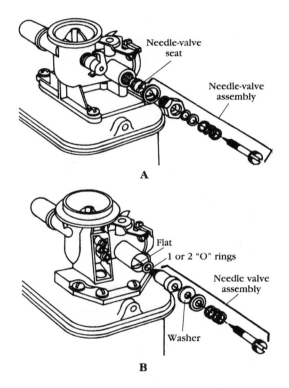

FIG. 4-24. *Two needle valve assemblies are used, one primarily associated with zinc carburetors (A); the other is used on all nylon-bodied models (B).* Briggs & Stratton Corp.

Figure 4-24 illustrates major Pulsa-Jet variations that closely track those of the Vacu-Jet. Most service information applies to both types.

Needle-valve assembly Sealing the needle valve, or mixture-adjustment screw, involves some fairly complex engineering. Figure 4-25A shows the

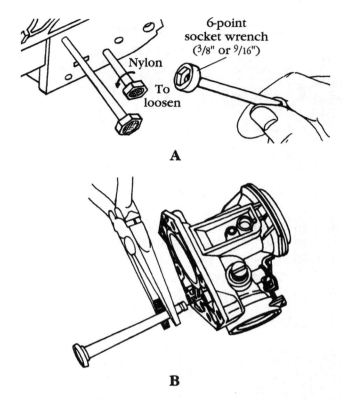

6-point
socket wrench
(³/8" or ⁹/16")

Nylon

To
loosen

A

B

FIG. 4-25. *Fuel pipes twist in and out of metallic carburetor castings (A). Pipes for nylon carburetors incorporate a snap lock (B).* Briggs & Stratton Corp.

arrangement of washers and O-rings generally found on pot-metal Vacu-and Pulsa-Jets. Figure 4-25B illustrates the arrangement always used on nylon carburetors and sometimes on the zinc models. The needle is quite vulnerable to damage from overtightening.

Pickup tubes Vacu-Jet fuel pickup tubes are fitted with a check ball, which tends to stick in the closed position. Because the ball and, on later models, the tube itself are made of nylon, more than a few minutes in carburetor cleaner is all that can be tolerated. As an emergency repair, you can free the ball by gently inserting a fine wire through the screen in the base of the tube. Eventually the assembly will have to be replaced.

Fuel pickup tubes supplied with zinc carburetors twist off and on. Tubes used with nylon carburetors snap in and out, an operation that can require considerable force (Fig. 4-25).

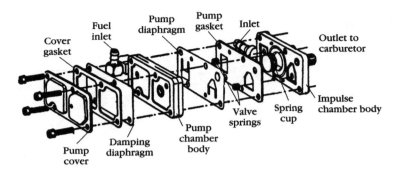

FIG. 4-26. *Diaphragm-type fuel pump.* Briggs & Stratton Corp.

Pulsa-Jet pump diaphragm The side-mounted diaphragm is shown in Fig. 4-26. The tank-mounted version, used with "bobtail" carburetors, is illustrated in Fig. 4-10. In either case, replace the diaphragm whenever the carburetor is serviced.

Fuel pump

Some engines are equipped with stand-alone versions of the crossover Flo-Jet pump. Fuel enters by gravity and leaves under pressure generated by a pulse-activated diaphragm. Replace the pump diaphragm as a routine part of fuel-system service.

Air filters

Briggs & Stratton supplies a variety of standard and optional air filters, which use foam or paper elements in combination with foam precleaners or "socks" (Fig. 4-27). The foam in early production filters tolerated kerosene and other petroleum-based solvents; the foam used on later designs does not. Wash all filter elements in warm water and nonsudsing detergent to avoid confusion and ruined filters. Rinse in the reverse direction of air flow until the water runs clear. Towel off the excess water and oil the element, kneading it as shown in Fig. 4-28. Foam elements should be reoiled as needed and whenever an engine that has been out of service for more than a few days is started. Oil tends to migrate out of the foam.

Pleated-paper elements, of whatever vintage, must not be exposed to oil or petroleum solvents. Some tolerate water. Wash in water and detergent. Allow the element to dry and the pores to shrink before installing. Do not blow out these fragile elements with compressed air.

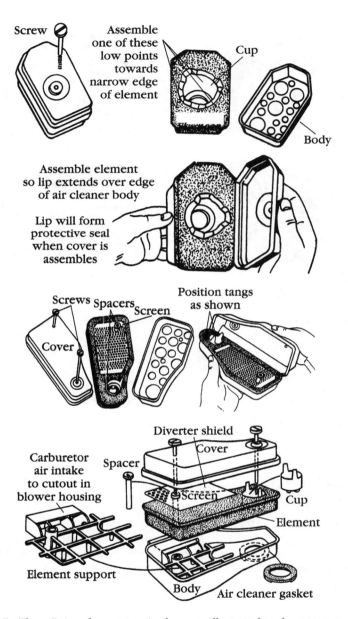

Screw

Assemble one of these low points towards narrow edge of element

Cup

Body

Assemble element so lip extends over edge of air cleaner body

Lip will form protective seal when cover is assembles

Screws Spacers Screen

Position tangs as shown

Cover

Diverter shield

Cover

Carburetor air intake to cutout in blower housing

Spacer

Screen

Cup

Element

Element support

Body

Air cleaner gasket

FIG. 4-27. *Three Briggs foam-type air cleaners, illustrated to show correct assembly procedures, which might not be obvious to the average mechanic. The more sophisticated paper-and-foam units offer less opportunity for error.*

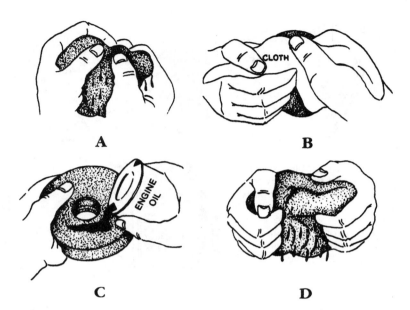

FIG. 4-28. *Wash the foam element in detergent and warm water (A). Wrap with a shop towel and press dry (B). Oil the element, wetting it thoroughly (C), and squeeze out the excess (D). Don't over-oil precleaners that stand against paper elements.* Briggs & Stratton Corp.

The cleaner-to-carburetor gasket should be renewed at the first sign of wear. Older and less expensive engines secure the filter with a single screw that, if bent, destroys the gasket seal. Replace the screw as necessary.

Governors

Small engine governors put a ceiling on no-load speed and hold rpm relatively constant under varying loads. Less expensive engines typically use air-vane governors of the general type illustrated in Fig. 4-29. The spring tends to close the throttle; the dynamic head of cooling air acting against the vane attempts to open the throttle. The manually operated throttle varies spring tension, relaxing it to allow the engine to run faster or, moved in the other direction, stretching the spring to slow the engine.

Centrifugal governors work in the same manner except that the closing force is generated by spinning weights, called *fly weights* (Figs. 4-30 and 4-31). These governors can be quite complex in detail and resist generalization.

Most governor malfunctions—hunting, lack of responsiveness, excessive no-load speed—can be corrected by replacing weak or distorted throttle

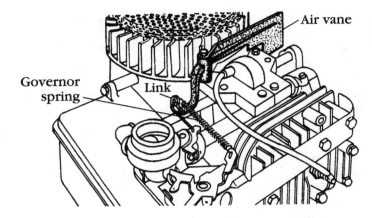

FIG. 4-29. *The typical Briggs & Stratton governor uses a plastic vane loosely secured with metal tabs.*

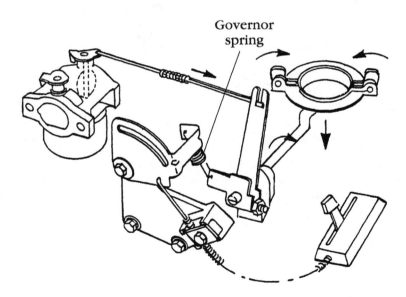

FIG. 4-30. *Centrifugal governor operation in response to load. Engine speed drops and the weights slow, allowing the spring to pull the throttle open wider. Details vary with engine model, but the pattern of forces—the spring pulls the throttle open, the weights tend to close it with a force proportional to engine rpm—applies to all.* Briggs & Stratton Corp.

springs and/or bent wire links. Do not change the geometry by installing springs or links in alternate mounting holes that might be present.

Warning: Governor springs are safety items that exert major influence on no-load speed. Ungoverned engines can act like grenades, exploding in fragments. Replace springs with the correct part number. Do not stretch or distort during installation. After any spring change, check no-load speed against the equipment manufacturer's specification with an accurate tachometer.

Mechanical governors incorporate an adjustment that, wrongly accomplished, affords the unwitting mechanic the opportunity to destroy the engine within seconds of startup. In most cases, the adjustment involves loosening the pinch bolt (Fig. 4-31) and rotating the shaft in a specified direction, relative to the lever. Because of the critical nature of this work, it should be farmed out to your local Briggs & Stratton distributor who, presumably, will stand good for mistakes.

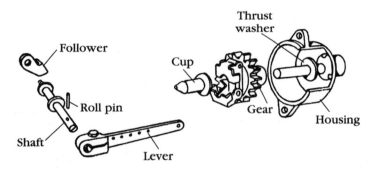

FIG. 4-31. *The governor mechanism used for 60000, 80000, and 140000 engines. The housing, accessed from outside of the engine, is unique to this engine family; all other centrifugal governors live inside of the crankcase. The pinch bolt that secures the lever and shaft is the main adjustment point for this and most other Briggs & Stratton governors and **should not be disturbed during normal service activities, including engine overhaul.***

5

Starters

Current Briggs & Stratton engines use a side-pull or vertical-pull rewind starter that might be supplemented with a dc- or ac-starter motor. Formerly, some of the company's products were fitted with a spring-powered unit, which is discussed briefly at the end of this section.

Rewind starters

The *rewind starter*, also known as the *recoil starter*, was introduced by Jacobsen in 1928 and has since become standard for small engines. While the constructional details differ, rewind starters have the following basic components:

- Pressed steel or aluminum housing, which locates the starter center over the flywheel center.
- Recoil spring, one end of which is anchored to the housing; the other to the pulley. The spring mounts in a recess in the pulley or housing.
- Nylon starter cord, sized by diameter and length to the application and secured to the pulley at one end and the handle at the other.
- Pulley, also known as the sheave ("shiv"), located relative to the starter housing by tabs or a bushing.
- Clutch, which transfers starter torque to the flywheel and automatically disengages when the engine starts. Briggs starters use either a sprag or friction clutch.

Troubleshooting

This section details some of the more common failures.

Broken rope   This is the most common failure and is often caused by the operator pulling too hard at the end of travel. Rope breakage might be abetted by a worn guide bushing (the ferrule that protects the rope from contact with the housing) or by a tendency of the engine to kick back during cranking.

Rope won't completely retract   Some spring action is present, which might be frustrated by starter-to-flywheel misalignment or by loss of preload. With the rope extended and the blower-housing bolts loosened a few turns, strike the housing with the palm of your hand. If the rope retracts, misalignment was the problem and the housing can be secured. If repositioning the housing has no effect, the recoil spring should be replaced. It is possible to salvage tired springs by increasing the preload, but at the cost of reduced rope travel.

No spring action   If the rope goes limp, the cause is almost certainly a broken spring.

Rope is hard to pull   Check the starter-to-flywheel alignment as described previously. Other possibilities include abnormal engine friction or, if rope resistance seems to pulsate, a loose flywheel or rotary mower blade. Check the sheave-axle bearing on Eaton starters.

Starter slips, and fails to engage the flywheel   This problem originates in the starter clutch and, on Eaton-pattern starters, almost certainly indicates a failed brake spring. Briggs clutches slip if extremely dirty.

Starter howls as engine runs   This is the classic symptom of a dry clutch/crankshaft bearing on Briggs-pattern starters. Remove the blower housing and apply a few drops of oil to the crankshaft end.

Eaton-pattern, side-pull starter

Quantum, Europa, and the better engines generally use versions of the Eaton starter. An Eaton starter employs spring-loaded clutch dogs that cam into engagement against the flywheel hub inside diameter (ID). The starter assembly is generally secured to the blower housing with pop rivets, although some engine models use bolts (Fig. 5-1). In any event, replacement starter assemblies bolt into place using the existing mounting holes.

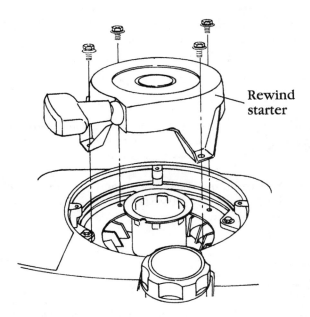

Rewind
starter

FIG. 5-1. *We will be seeing more Eaton starters as Briggs turns to outside vendors for components. While these starters are not identical to those used by other manufacturers, clutch parts can be purchased from Tecumseh dealers. This illustration does not show the plastic grass guard that mounts over the starter housing.*

Disarming

First release spring preload tension, which on Eaton starters is done by removing the handle and allowing the sheave to unwind in a controlled fashion. Brake the sheave with your thumbs. It is also helpful to count the number of sheave rotations from the point of full-rope retraction so that the same preload can be established upon assembly.

Warning: Even after preload has been dissipated, the spring remains confined in its housing under considerable tension. Wear safety glasses when servicing these starters.

Rope replacement

If the rope has parted, preload has already been lost. If the rope remains in one piece, untie the knot securing the handle. Allow the rope to fully retract, braking the sheave as described in the preceding paragraph.

Briggs & Stratton supplies precut and fused starter cords for the various engine models. If you purchase stranded nylon cord in bulk, replicate the original diameter ($\#4\frac{1}{2}$ for 60000 through 120000; $\#5\frac{1}{2}$ for 130000 and larger) and length. Fuse the cut ends in an open flame.

Warning: Fused rope ends retain enough heat to produce painful burns for several minutes after exposure to flame.

Follow this procedure:

1. Tie a square knot in one end of the rope (Fig. 5-2).

FIG. 5-2. *A figure-eight knot secures the rope to the sheave. Rope ends should be melted to prevent unraveling.* Briggs & Stratton Corp.

2. Using a screwdriver in the sheave slot, wind the spring clockwise (as viewed from the underside of the starter) until tight.
3. Allow the sheave to unwind enough to align the slot with the rope eyelet (Fig. 5-3).

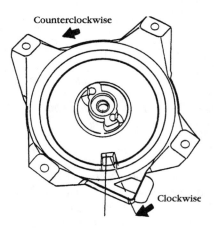

FIG. 5-3. *Turn the sheave counterclockwise to coil bind and back off far enough to align the rope cavity with the eyelet. This establishes preload for the Briggs-Eaton starters discussed here. If you are dealing with an unfamiliar starter, preload is the number of revolutions the sheave made after "swallowing" the rope. If after assembly, the rope fails to retract smartly, add a revolution or so of preload. If the spring coil binds near the end of rope travel, release some of the preload.*

4. Insert the unknotted end of the rope into the slot and through the eyelet (Fig. 5-4).
5. Tie a temporary knot in the rope or lock the sheave with Vise-Grip pliers (Fig. 5-5).
6. Install the handle and secure with a knot. Release the sheave and test starter action.

Clutch

Two spring-loaded dogs inside a pressed aluminum retainer transmit starting torque to the flywheel hub. A compression pin or screw secures the retainer to the underside of the sheave and the sheave to its axle. Note that some retainer screws have left-hand threads.

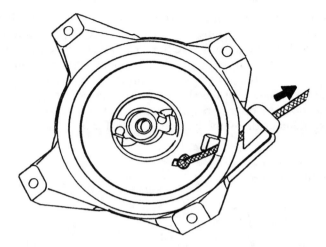

FIG. 5-4. *Thread the rope through the eyelet, or bushing, and seat the knot in the sheave cavity. To avoid fighting the spring while tying on the handle, clamp the sheave to the housing with Vise-Grips.* Briggs & Stratton Corp.

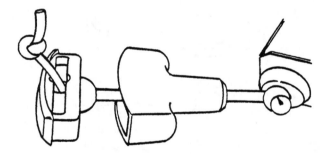

FIG. 5-5. *Knot the rope as shown at the handle.* Briggs & Stratton Corp.

Warning: Wear eye protection when servicing rewind starters, particularly during and subsequent to removal of the retainer fastener. All that holds the main spring captive is a shallow recess in the upper side of the sheave.

Engine model	Clutch retainer fastener	Torque
95700, 96700 (2-cyc)	Hex-head, LH thread	30 lb./in.
99700 (Europa)	Pin	
104700 (OHV)	Phillips, RH thread	70 lb./in.
120000 (Quantum)	Pin	

Figures 5-6A and 5-6B illustrate these arrangements. Note that the fastener preloads the brake spring against the retainer. Spring tension causes the retainer and sheave to turn together as an assembly during the first few degrees of sheave rotation. Further movement of the retainer is then blocked by the dogs that, at this point, grip the flywheel hub in full extension. Continued rotation of the sheave exerts a steady drag on the retainer to hold the dogs in engagement. When the starter cord is released, the sheave rewinds and drags the retainer with it for a fraction of a turn, just

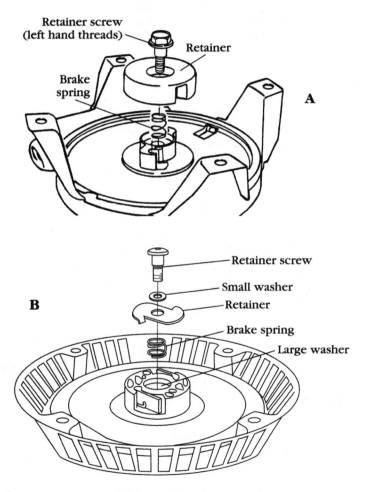

FIG. 5-6. *Eaton starters are held together with a screw (A) or throwaway compression pin (B). Washer sequence varies with starter model but all employ a coil spring as a friction generator.* Briggs & Stratton Corp.

enough to back out the dogs and release the flywheel. In practical translation, this means:

- Brake springs are the most vulnerable part of the assembly.
- A loose retainer pin or screw dissipates brake spring preload, resulting in starter slip.
- Retainers undergo severe wear at the brake spring and dog contact points.
- Bent or distorted dog springs should be replaced.
- Lubrication is an enemy.

Use a punch to drive out the compression pin from the out-board side of the starter housing to the engine side. In some cases, a decorative decal must be peeled from the starter housing to gain access to the pin. Support the housing and sheave with PN 19227 or a short length of pipe placed vertically beneath it (Fig. 5-7). From the engine side, hammer or press in a new replacement pin and install spring and washers. Seat at the original depth.

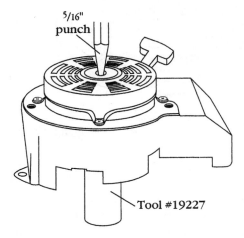

FIG. 5-7. *PN 19227 is a hollow cylinder that allows the pin to drop while keeping the sheave on its axle and the main spring contained. A piece of 3-in. pipe can substitute.* Briggs & Stratton Corp.

Spring & sheave

The spring lives in a recess in the sheave between the sheave and underside of the starter housing. One end of the spring anchors to the sheave, the other to the housing.

Springs used for engine Models 104700 overhead-valve (OHV) and 120000 (Quantum) are considered an integral part of the sheave and should not be separated from it.

99700 and more recent models have replaceable springs. Grasp the spring with long-nosed pliers and carefully lift it out of the sheave (Fig. 5-8) and release it inside of the starter housing (Fig. 5-9), which acts as a kind of cage.

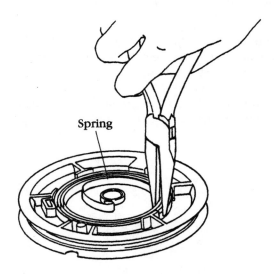

Spring

FIG. 5-8. *Grasp the spring with long-nosed pliers and carefully lift it out of the sheave recess on engine models with replaceable springs.* Briggs & Stratton Corp.

Warning: Wear safety glasses and long sleeves and gloves.

Inspect the sheave for cracks in the hub and damage to the spring anchor. Plastic sheaths require no lubrication, although a bit of oil on the axle helps prevent rust.

Replacement springs are packaged in a plastic retainer for ease of handling. A small dab of grease is all the lubrication required. Insert the outer spring end in the slot provided and, holding the coils with long-nosed pliers, cut the retainer loose (Fig. 5-10). Reinstalling the original spring, which does not have a retainer, is a matter of anchoring the outer end and laying the spring down counterclockwise, a coil at a time (Fig. 5-11). Install the sheave on the axle and rotate it counterclockwise to engage the anchor.

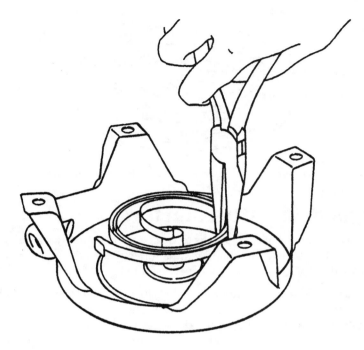

FIG. 5-9. *Release the spring inside of the starter housing.* Briggs & Stratton Corp.

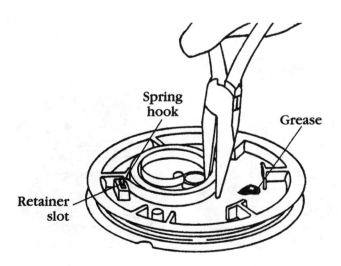

FIG. 5-10. *New springs are first anchored and then released by cutting the retainer band.* Briggs & Stratton Corp.

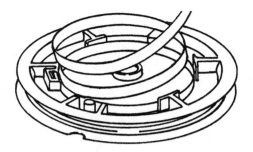

FIG. 5-11. *Used springs can be wound in by hand.* Briggs & Stratton Corp.

Briggs & Stratton starters

The Briggs side-pull starter continues to be specified for many single-cylinder models. Unlike other rewind starters, it is integral with the blower housing and drives through a sprag, or rachet-type, clutch.

Sprag clutch   Recoil and impulse starters drive through a sprag clutch that doubles as the flywheel nut. The clutch housing (Fig. 5-12) threads over the crankshaft. The sprag (ratchet in the drawing) is supported by a bushing on the crankshaft stub. Its outside end mates with the starter pulley, and its lower, or inside, end rides against four or six ball bearings in the starter housing. When rotated by the starter pulley, the sprag traps a ball bearing between it and the clutch housing, locking the starter to the crankshaft. Once the engine catches, the ball bearing releases and the sprag idles on the bushing.

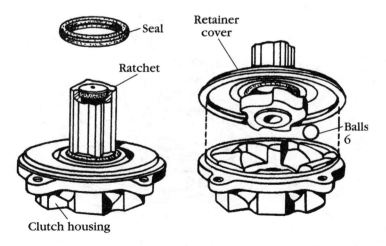

FIG. 5-12. *The Current production sprag clutch.* International Harvester Corp.

To service the clutch, remove the engine shroud and the screen, which is mounted to the clutch housing by four self-threading screws. Disconnect and ground the spark-plug lead to prevent accidental starting. Secure the flywheel with a strap wrench or a Briggs & Stratton holding fixture. Unthread the clutch assembly using factory tool PN 19161 or 19114. If this tool is not available, the assembly can be loosened with a hammer and a block of soft wood. Some damage to the screen lugs is inevitable but is less than fatal if distributed evenly to all four lugs. A spring washer fits under the clutch assembly.

On early models, the retainer cover was secured with a spring wire; on late models, the cover must be pried off. Clean the sprag, clutch housing, and ball bearings in solvent. Some deformation of the clutch housing is normal. Wear on the tip of the sprag, the part that makes contact with the bearings, can cause the clutch to slip. Reassemble these parts dry, without lubricant, and lightly oil the bushing. Install the spring washer and torque to specifications in Table 5-1.

Table 5-1
Clutch housing torque limits

Cast-iron series	Torque (ft-lb)
6B, 6000, 8B, 80000, 82000, 92000, 110000	55
100000, 130000	60
140000, 170000, 1717000, 190000, 191700, 251000	65
Aluminum series	
19, 190000, 200000	115

Horizontal pull starter To dismantle the starter, remove the shroud and place the assembly upside down on a bench (Fig. 5-13). Cut the rope at the sheave knot and extract it. Use a pair of pliers to pull the main spring out of the housing as far as it will come (Fig. 5-14). The purpose is to bind the spring so that it will not "explode" when the sheave is detached. For further protection, wear safety glasses. Carefully straighten the sheave tangs. Withdraw the sheave, twisting it slightly to disengage the spring. Clean the metal parts in solvent and inspect for damage.

Secure the blower shroud to the workbench with several large nails or hold the shroud lightly in a vise. Lightly grease the spring and attach it to the sheave. Thread the free end out through the anchor slot in the shroud. The hole in the sheave measures $3/4$ in. square. A 6 in. length of a 1 × 1 or the male end of a $3/4$-in. drive-extension bar can be used to wind the sheave. With the tangs bent down into light rubbing contact, rotate the sheave and

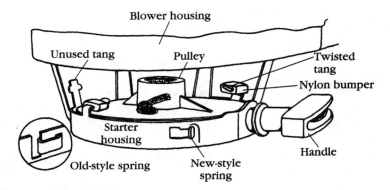

FIG. 5-13. *Some models employ an insert at the housing spring-anchor slot.* Briggs & Stratton Corp.

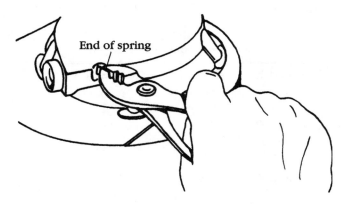

FIG. 5-14. *Disarm the spring before disassembly.* Briggs & Stratton Corp.

wind the spring tight. With your free hand, guide the spring through the slot in the shroud. Press the notched end of the spring into engagement with the anchor slot.

Without releasing the sheave, thread the rope into it. A length of piano wire can be used as a pilot. Fish the end of the rope through the knot hole, tie it, and seal the frayed edges with a match. Push the knot down into the hole for clearance. The process is the same with new-style pulleys except there is no lug to frustrate your work.

Secure the handle with a figure eight knot, leaving about ¾ in. of rope beyond the knot. Seal the end with heat and slip the handle pin through one of the knot loops.

Release the spring in a controlled manner and allow the rope to wind. Bend the lugs so that the nylon bumpers are 1/16 in. below the sheave (the bumpers were against the sheave during winding for better control). Install the shroud assembly on the engine, centering it over the flywheel. Test the starter. If it is slow to retract or binds, loosen the shroud and reposition it.

Vertical pull starter   The vertical pull starter is a convenience on vertical crankshaft engines because it eliminates the need to crouch alongside the engine to start it. This starter is considered a safety feature on rotary lawn-mowers. Pulling on the rope sends a nylon gear into engagement with the ring gear on the underside of the flywheel. The nylon gear moves on a thread by virtue of a friction spring and link (an arrangement reminiscent of that used on bicycle coaster brakes). Once the engine fires, the gear retracts back down the thread.

Warning: Wear safety glasses when servicing this and other starters.

The main spring is under some tension. Disarm the spring by lifting the rope out of the sheave groove and winding the sheave, together with the freed section of rope, several turns counterclockwise (Fig. 5-15). When you are finished, there should be no tension on the sheave and approximately 12 in. of rope should be free. Observe the warning stamped on the plastic starter cover and, using a screwdriver, gently pry the cover off. Do not pull on the rope with the cover disengaged.

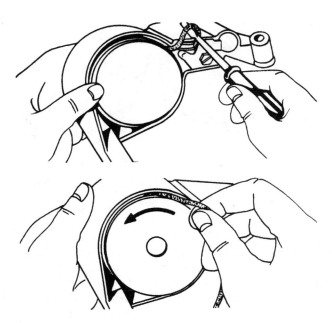

FIG. 5-15. *Disarm the Briggs vertical pull starter by lifting a foot or so of rope out of the pulley groove.*

Remove the anchor bolt and anchor and note how the spring mates with it (Fig. 5-16). If the spring is to be replaced, carefully work it out of the housing. Remove the rope guide and observe the position of the link (Fig. 5-17) for assembly reference. Using a piece of piano wire in conjunction with long-nosed pliers, pull the rope far enough out of the sheave to cut the knot. Clean mechanical parts in solvent. The friction spring and link are the most vulnerable elements in this mechanism. See that the link and spring assembly move the drive gear to its extremes of travel. If there is any hesitation, replace these parts.

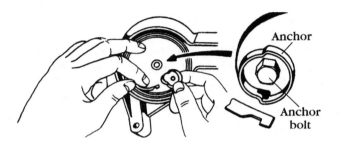

FIG. 5-16. *Remove the anchor bolt and spring.* Briggs & Stratton Corp.

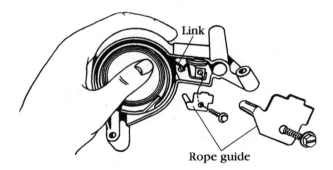

FIG. 5-17. *Observe the position of the friction link before disassembly.* Briggs & Stratton Corp.

Begin reassembly by installing the spring in its housing. Slip one end into the retainer slot and wind the spring counterclockwise (Fig. 5-18). Using a length of piano wire or a jeweler's screwdriver, snake one end of the rope into the pulley. Extract the end of the rope from behind the sheave and tie a small, hard knot. Space is critical. No more than 1/16 of an inch of rope should extend beyond the knot. Melt the ends with a flame and wipe down the melted fibers with a shop rag to reduce their diameter. Pull the rope tight and check that the knot clears the threads.

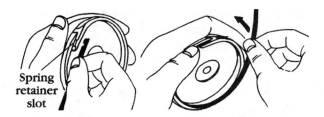

FIG. 5-18. *To install the spring, anchor it in the retainer slot and wind counter-clockwise.* Briggs & Stratton Corp.

Install the rope guide with the link positioned as it was originally found (Fig. 5-19). Wind the spring counterclockwise with your thumbs to retract the rope (Fig. 5-20). Once the handle butts against the starter case, secure the spring anchor with 80 to 90 lb/in. of torque. Lightly lubricate the spring with motor oil.

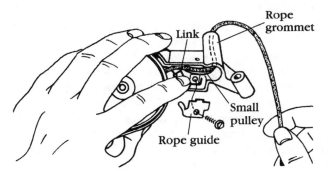

FIG. 5-19. *Install the friction link behind the rope guide.* Briggs & Stratton Corp.

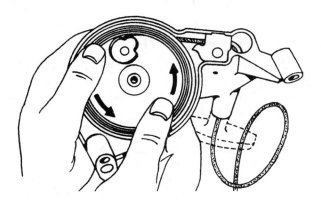

FIG. 5-20. *Wind the spring counterclockwise to retract the rope.* Briggs & Stratton Corp.

Snap the starter cover into place and disengage approximately 12 in. of rope from the sheave. Give the rope and sheave two or three clockwise turns to preload the main spring and to assure that the rope will be rewound smartly (Fig. 5- 21).

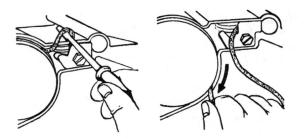

FIG. 5-21. *Preload the spring two or three rotations.* Briggs & Stratton Corp.

Impulse starters

The spring-powered impulse starter should not be given new life from repairs. These devices were inefficient, generally unreliable, and quite dangerous. In most cases, a B & S side-pull rewind starter can be easily substituted. However, the impulse starter must be at least provisionally disarmed.

Release tension by placing the control knob or remote-control lever in the "start" position. If the engine is locked and the starter fails to unwind, turn the knob or lever to the "crank" position. Get a strong grip on the crank handle with one hand and remove the Phillips screw at the top of the assembly.

Warning: If unsecured, the crank handle can spin with enough force to break an arm.

Dispose of the old starter without dismantling it further or handling the parts more than necessary. The main spring retains a serious potential for bodily damage.

Starter motors—various models

In recent years, Briggs has used a variety of starter motors, manufactured internally and purchased from second parties. Both 12 Vdc and 120 Vac models are supplied. The following discussion covers the more popular models but is not all-inclusive.

Does not crank	low battery; low line voltage (120 Vac); high resistance connection; open starter switch; heavy load; defective motor or rectifier (120 Vac)
Cranks slowly	low battery; low line voltage (120 Vac); high resistance connection; worn motor bearings; worn or sticking brushes; heavy load

Figure 5-22 illustrates three starter motors available for the 140000-, 170000-, and 190000-series engines. These motors are typical of all gear-driven types of motors.

Briggs & Stratton suggests two test parameters, no-load rpm, and no-load current draw, for the motors that dealer mechanics service. To perform these tests, you will need a hand-held tachometer, an ammeter, and a power supply. Depending on the starter motor, the power supply is a fully charged 6 V or 12 V lead-acid battery or 12 V Nicad battery, or a 120 Vac source. The current readings in Table 5-2 are steady draw readings—disregard initial surges.

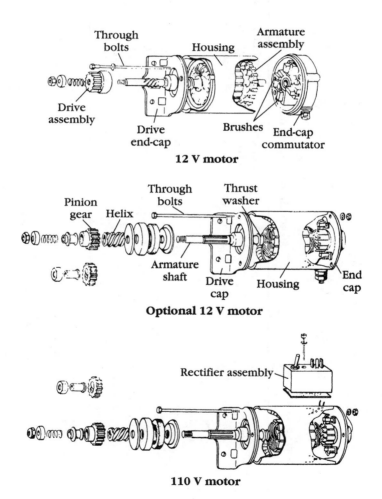

12 V motor

Optional 12 V motor

110 V motor

FIG. 5-22. *Typical starter motors.* Briggs & Stratton Corp.

Table 5-2
Steady draw current ratings

Motor	Engine model	Minimum (rpm)	Maximum current draw	Power supply
12 Vdc geared	140000, 170000, 190000	5000	25.0 A	6 V battery
12 Vdc geared (American Bosch No. SMH 12A11)	140000, 170000, 190000	4800	16.0 A	12 V battery
110 Vac geared	140000, 170000, 190000	5200	3.5 A	110 Vac
12 Vdc geared	130000	5600	6.0 A	12 V battery
110 Vac geared	130000	8300	1.5 A	110 Vac
12 Vac geared	300400, 320400	5500	60.0 A	12 V battery
12 Vdc geared (Nicad)	92000, 110900	1000	3.5 A	12 V Nicad battery

Mark the end cap and motor frame for assembly reference and remove the two through-bolts that secure the cap to the frame. Take off the brush cover and cap. The armature can be withdrawn from the drive side with the pulley still attached. Starter motor failure can be traced to:

- binding (scored or dry) armature shaft bearings
- worn armature shaft bearings
- shorted, opened, or grounded armature
- shorted, opened, or grounded field
- brushes worn to half or less of their original length
- brushes sticking in their holders.

Reddish brown discolorations on the commutator bars are normal and mean that the brushes have seated. Burned commutator bars signal a shorted winding. Glaze and minor imperfections can be removed with number 00 sandpaper as shown in Fig. 5-23. Severe out-of-round, deep pits, or scores should be corrected with a lathe. After any of these operations, cut

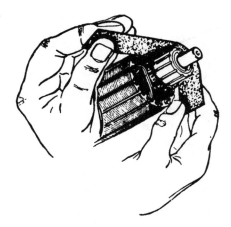

FIG. 5-23. *Cleaning the commutator.*
Tecumseh Products Co.

down the mica with a tool designed for this purpose, or with a narrow, flat-edged jeweler's file (Fig. 5-24). Polish the commutator to remove burrs and clear the filings with compressed air.

Bearings are the next most likely area of failure. The starter might turn freely by hand, but when engaged against the flywheel groan through a revolution or so, and then bind.

Drive out the old bushings, being careful not to score the bearing bosses, and drive in new ones to the depth of the originals. Bushings in motor end covers can be removed by any of several methods. A small chisel can be used to split the bushing. American Bosch end-cover bushings have a flange

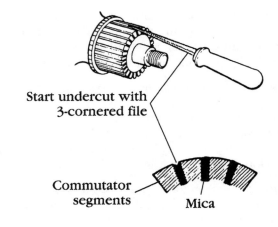

Start undercut with
3-cornered file

Commutator
segments

Mica

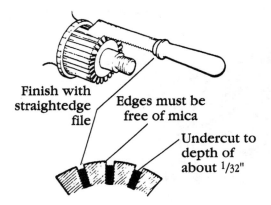

Finish with
straightedge
file

Edges must be
free of mica

Undercut to
depth of
about $1/32$"

FIG. 5-24. *Undercutting the mica.* Kohler of Kohler.

to accept thrust loads that can be used as a purchase point to collapse the bushing inward. The neatest technique is to pack the boss with heavy grease, then ram the bushing out with a punch that matches the diameter of the motor shaft. A sharp hammer blow will lift the bushing by hydraulic pressure.

The armature can develop shorts. Check for shorts between the shaft and armature with a 120 Vac test lamp. All iron and steel parts must be electrically isolated from nonferrous (brass or copper) parts. Check adjacent commutator bars by the same method. Handle the 120 Vac probes with extreme caution—holding one in each hand means that an electric current could pass through your vulnerable thorax.

Internal winding-to-winding shorts can be detected with a growler. These tools are fairly expensive to buy, but a few autoparts houses keep one for customer use. You can build one around the core of a television power supply transformer.

If one of the armature windings is shorted, a hacksaw blade will vibrate when placed over the affected armature segment (Fig. 5-25). An open winding will generate sparks between the blade and adjacent commutator segments. Unless the damage is visible, such as a broken connection between the armature and a commutator bar, there is no practical way to repair an armature. Rewinding costs more than a replacement.

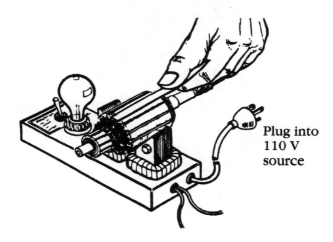

Plug into
110 V
source

FIG. 5-25. *Checking for internal (winding-to-winding) shorts with a growler and hacksaw blade.* Tecumseh Products Co.

American Bosch motors use permanent magnetic fields that require no service under normal circumstances. Arc welding on adjacent parts or extreme vibration can weaken the magnets. Few shops have the necessary equipment to "recharge" magnets and the fields must be replaced.

Brushes must be at least half their original length to maintain pressure against the commutator bars. One brush or brush set should be grounded to the frame, while the remaining brush or brush set connects to the armature. The brushes should be free to move in their holders, and the assembly must be free of carbon dust and oil deposits. In most cases, the brushes must be "shoe-horned" over the commutator with a homemade tool (Fig. 5-26).

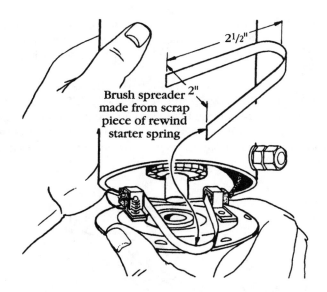

FIG. 5-26. *A homemade tool used to overcome brush-spring tension during reassembly.* Briggs & Stratton Corp.

6

Charging systems

Engine models employ more than a dozen distinct alternators for use with lead-acid batteries. The Nicad system is recharged from house current through a step-down transformer and rectifier.

Storage batteries

Storage batteries can fail mechanically or electrically. The leading causes of mechanical failure are loose battery hold-down hardware, poor vibration insulation, and owner abuse. The battery straps—the internal busses that connect the cells—are cast as part of the terminals. Twisting the cables or overtightening the terminal bolts can fracture the straps.

Electrical failure is usually associated with chronic low states of charge. The plates become impregnated with sulphate crystals and are no longer capable to take part in the ion exchange that generates electrical potential. A partial cure is to trickle charge the battery for a week or more. Some of the sulphate dissolves into solution. However, the best cure is prevention. Distilled water should be added to cover the plates with electrolyte, and the state of charge should be held greater than 75 percent, or 1.1220 on a temperature-corrected hydrometer.

Deep-charge/discharge cycles encourage sulphation and, if the system is not properly regulated, can overheat the battery and melt holes in the plates. The extent and rate of discharge can be reduced by keeping the battery charged and the engine in tune. The less cranking the better, particularly if the battery shows signs of fatigue. Self-discharge can be controlled by frequent transfusions and by keeping the battery top and terminals clean. The rate of charge is, practically speaking, beyond owner control, although it is wise to invest in an ammeter to keep an eye on the system.

Before turning to specific test procedures, it should be noted that the capacity of the battery has some bearing on its longevity. All things being equal, a larger battery will outlive one that delivers its last erg of energy each time the engine is cranked. But the capacity of the battery, usually measured in ampere hours (A or Amp), cannot compensate for long-term withdrawals. Ultimately, even the large battery must be recharged depending on the output of the alternator and the way the engine is used. A small alternator that is adequate for one start a day might not deliver the current for 20 starts a day.

The first evidence of charging-system failure is a low battery. The state of charge—how much potential is available in the battery—is easily measured with a hydrometer. While hydrometer results do not take the place of a performance test, the hydrometer is the instrument to be tried first.

A hydrometer consists of a squeeze bulb, a float chamber, and a precisely weighted float (Fig. 6-1A). The float is calibrated in units of specific gravity. Water has a specific gravity of 1.000. Sulphuric acid, the other ingredient of electrolyte, has a specific gravity of 1.830. In other words, sulphuric acid is 1.830 times heavier than an equal amount of water. The amount of acid in the electrolyte reflects the state of charge. The more acid in the electrolyte, the greater the charge and the heavier the electrolyte. Each cell in a fully

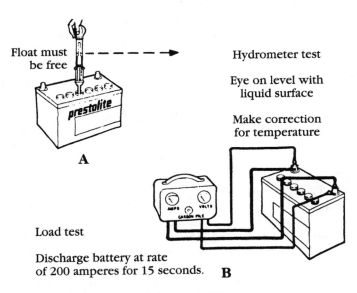

FIG. 6-1. *A battery hydrometer and rheostat OMC.*

charged battery should have a specific gravity of 1.240–1.280. A completely discharged battery will have a specific gravity of about 1.100.

Draw enough electrolyte into the hydrometer to set the float adrift. The float must not touch the sides of the instrument. Sight across the main level of the instrument. Disregard the meniscus that clings to the sides of the chamber and record the specific gravity for that cell. Repeat the operation on the other cells. The battery should be suspected of malfunctioning if any of the cells fall five points (0.005) below the average of the others.

While raw, uncorrected readings are generally adequate, it should be remembered that acid and water expand when heated. The higher the temperature of the electrolyte, the lower the apparent specific gravity. Expensive hydrometers sometimes incorporate a thermometer in the barrel and a temperature-compensating scale. Any accurate thermometer will work. For each 10 degrees above 80 degrees Fahrenheit add four points (0.004) to the reading. Subtract four points for each 10 degrees less than 80 degrees Fahrenheit.

The most reliable field test requires a carbon pile (Fig. 6-1B) or a rheostat and a voltmeter. The temperature-compensated specific gravity should be at least 1.220 to prevent battery damage. Connect the voltmeter across the terminals and adjust the load to three times the ampere-hour rating. For example, the carbon pile should be adjusted to discharge a 30 A battery at the rate of 90 amps. Continue to discharge for 15 seconds. At no time during the test should the voltmeter register less than 9.6 V. If it does, the battery should be suspected of malfunctioning.

Alternators

Briggs & Stratton provides battery-charging current with an engine-driven alternator and solid-state rectifier. The rectifier converts alternating current into pulsating direct current. The more sophisticated systems include a solid-state voltage regulator to protect the battery from overcharging and to extend headlamp life. Optional features include an ammeter and an isolation diode. All systems employ a lead-acid storage battery.

System 3 & System 4 alternators

Model 90000 and 110000 engines currently use System 3 and 4 alternators, which employ two coils and a full-wave rectifier on a dedicated stator. Versions available are 6 Vdc and 12 Vdc. Output should be 0.5 A at 2800 rpm, as measured between the battery lead and engine ground. Replace the alternator if the output is under specification. Set the stator-to-flywheel air gap for ignition coils at 0.010 in., as described in chapter 3.

The 0.5 A alternator

The 0.5 A alternator is the peewee of the series, consisting of a single charging coil and integral rectifier. Found on model 120000 vertical and horizontal crankshaft engines, the unit should deliver 0.5 A at 2800 rpm. If it fails to produce its rated output, replace the assembly. The early version used a pinch bolt, since discarded, for air gap adjustment, which is set at 0.07 in.

The 1.5 A alternator

The three-coil 1.5 A alternator is used on 130000 series engines (Fig. 6-2) where it is coupled to a 12 A or 24 A battery. Note that there are two styles of connector, both of which contain a soldered-in rectifier, available as a replacement part. Other than the resistance provided by the battery, the circuit has no voltage or current regulation.

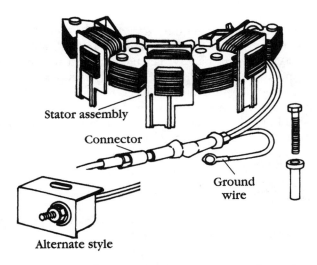

Stator assembly

Connector

Ground wire

Alternate style

FIG. 6-2. *1.5 A alternator found on 130000 engines. Note the variance in rectifier type.* Briggs & Stratton Corp.

To test the alternator output, connect a number 4001 headlamp between the rectifier output and a paint-free engine ground (Fig. 6-3). The battery must be out of the circuit. Under no circumstances should the output of this or any other alternator be grounded. To do so is to invite burned coils and fried diodes.

If the lamp refuses to light, the fault is in the rectifier or the alternator. Test the rectifier first because it is the more likely failure point. With the engine stopped, touch the probes of a low-voltage ohmmeter to the output

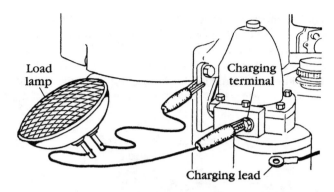

FIG. 6-3. *Testing output of the 1.5 A alternator.* Briggs & Stratton Corp.

terminal and ground as shown in Fig. 6-4. You should get continuity in one direction and high resistance in the other. If not, replace the rectifier box.

Test the stator with a 4001 headlamp connected across the output leads (Fig. 6-4). The lamp should burn. If not, check the leads to the stator for possible fouling. Before deciding that the stator is defective, compare the magnetic strength of the flywheel ring against one known to be good. Failure is exceedingly rare but not impossible.

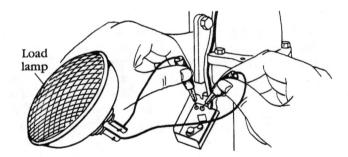

FIG. 6-4. *Testing stator of the 1.5 A alternator.* Briggs & Stratton Corp.

Install a replacement stator and torque the cap screws 18 to 24 lb/in. See that the output leads are snug against the block and well clear of the flywheel.

Dc-only & ac-only alternators

The dc-only alternator is found on 170000 and larger engines; the ac version on 190000 and larger engines. An identical four-coil stator produces 5 A for the ac version and, because of rectifier resistance, 4 A for the dc unit

(Fig. 6-5). The ac alternator should produce 14 Vac no-load at 3600 rpm. If not, replace the stator. Current output from the dc version, as measured by an ammeter in series with the charging lead, should range between 2 A and 4 A, depending on battery voltage. If no or low output is indicated, test the in-line diode with an ohmmeter. It should register low resistance in one direction and high resistance in the other. If the diode tests okay, the stator is at fault.

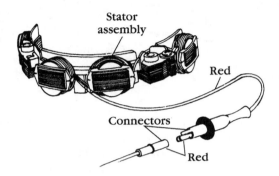

FIG. 6-5. *The dc-only rectifier can be distinguished from its ac-only twin by the bulge in the connector, signaling the presence of a diode rectifier.* Briggs & Stratton Corp.

The 4 A alternator

Used on 17 and 190000 engines, the 4 A alternator has eight charging coils arrayed on a 360 degree stator. This alternator does not include a regulator, but is otherwise identical to the 7 A type illustrated in Fig. 6-5.

Troubleshooting procedures begin with a short-circuit test. Connect a 12 V test lamp between the rectifier output and the positive terminal of a charged battery (Fig. 6-6). If the lamp lights, battery current is being fed back to ground through the charging circuit. Unplug the rectifier connection under the blower housing. If the lamp goes out, the rectifier is okay and the problem lies in the alternator and associated wiring. If the lamp continues to burn, the rectifier is at fault and must be replaced.

Inspect the output leads from the alternator for frayed insulation or other evidence of shorts before you replace the stator assembly. Make necessary repairs with electrical tape and silicone cement.

This alternator has four distinct windings, each involving two coils. A break in one of the windings drops output by a third. Check each of the four pins with the fuse-holder lead as shown in Fig. 6-7. Each pair of pins supplies current to a diode in the rectifier. Should a diode blow, a quarter of the output is lost. Check each of the four rectifier terminals with an ohmmeter. One probe should be on a good (paint-free and rust-free) ground on the

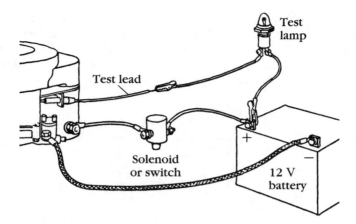

FIG. 6-6. *Testing for shorts in the 4 A, 7 A, and dual-circuit alternators.* Briggs & Stratton Corp.

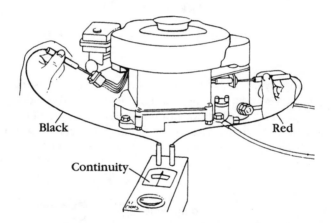

FIG. 6-7. *Testing the stator in the 4 A and 7 A alternator.* Briggs & Stratton Corp.

underside of the blower housing; the other probe should be on one of the diode connection points. Observe the meter and reverse the test connections. If the diode in question is functional, it will have continuity in one direction and very high resistance in the other. Repeat the test for the remaining three connection points.

The 7.0 A alternator

The 7.0 A alternator is used on series 140000, 170000, and 190000 engines and can be easily recognized by the connector plug, which is flanked by a

regulator on one side and the rectifier on the other (Fig. 6-8). Some instal-
lations employ an isolation diode in a tubular jacket on the outside of the
shroud. The purpose is to block current leakage from the battery to ground
by way of the alternator windings. Applications that do not have this diode
isolate the battery at the ignition switch.

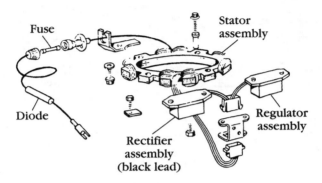

FIG. 6-8. *The 7.0 A alternator configuration.* Briggs & Stratton Corp.

Test the isolation diode by connecting a 12 V lamp in series with the
output (Fig. 6-9). The lamp should not light. If it does, the diode is shorted
and must be replaced. Check diode continuity with an ohmmeter connected
between the two diode leads. The meter should show zero resistance in one
direction and high resistance when the leads are reversed.

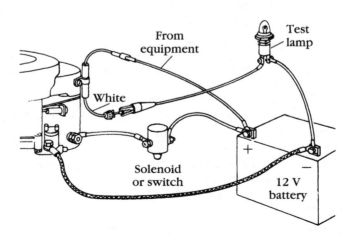

FIG. 6-9. *Testing the isolation diode on the 7.0 A alternator.* Briggs & Stratton Corp.

To test the stator, regulator, or rectifier, connect a test lamp as shown previously in Fig. 6-6. Do not start the engine. If the lamp lights, one of the three is shorted. Disconnect the rectifier-regulator plug under the blower housing and remove the stator from the circuit. If the lamp continues to burn, the regulator or rectifier is shorted. Test these two components individually to determine which is at fault.

Test the rectifier as described previously. Two black leads, joined by a connector, go to the rectifier. Each lead services two pins on the rectifier side of the connector. Without removing the rectifier assembly from the shroud, connect ohmmeter leads between each of the four pins and a paint free ground on the underside of the shroud. Observe the meter reading at each pin and reverse the leads. The pins should conduct in one direction and not in the other. If current flows in both directions, the rectifier is shorted. If no current passes, the rectifier is open. In either event, the assembly must be replaced. Instructions are packaged with the new part.

The regulator is distinguished by one red and one white lead. Test as above. The white lead pin must show some conductivity in one direction and none, or almost none, in the other. The red lead pin should give no reading in either direction. If it is necessary to replace the regulator, instructions are supplied with the replacement part.

Check stator continuity as shown previously in Fig. 6-7. Each of the four pins must be contiguous with the lead at the fuse holder. If not, check the visible wiring for defects before you invest in a new stator.

The dual-circuit alternator   The dual-circuit alternator is one of the most interesting alternators used on Briggs & Stratton engines. Two stator windings are provided, one for the headlights and the other for the battery. Battery output is rectified and rated at 3 A. Headlight output is alternating and can deliver 5.8 A at 12 V at wide-open throttle. Two versions of this alternator, one with a single plug and the other with separate ac and dc output lines, are used on 170000 and Lorgen engines.

The battery circuit is protected by a 7.5 A type AGC or 3AG automotive fuse and might be supplied with an ammeter. The ac circuit is independent of the charging circuit, although good practice demands that both be grounded at the same engine mounting bolt. Each circuit is treated separately here.

Charging circuit   Check output with an ammeter in series with the positive battery terminal (Fig. 6-10). The meter should show some output at medium and high engine speeds. No charge indicates a blown fuse, shorted or open wiring, or a failed rectifier or alternator.

Connect a 12 V test lamp between the battery and the charging section as illustrated in Fig. 6-6. The lamp should not light. If it does, the alternator

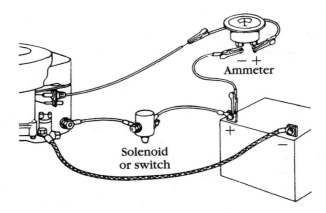

FIG. 6-10. *Testing dc output on the dual-circuit alternator.* Briggs & Stratton Corp.

or rectifier is defective. To isolate the problem, disconnect the plug under the shroud. If the light goes out, the rectifier is good and the difficulty is in the alternator or its external circuitry. If the light continues to burn, the rectifier is shorted to ground and must be replaced.

To test the stator, remove the starter motor, shroud, and flywheel. Inspect the red output lead for frayed or broken insulation. Repair with electrical tape and silicone, being careful to route the lead away from moving parts. Test the stator by connecting an ohmmeter between the terminal at the fuse holder and the red lead pin in the connector. The meter should show continuity. If not, the stator is open and must be replaced.

Test for shorts by connecting the ohmmeter between a good ground and each of the three black lead pins in sequence (Fig. 6-11). Test for continuity by holding the probes against the two black pins (Fig. 6-12). If the circuit is open, the stator is good. If the meter shows continuity, the stator must be replaced.

The rectifier mounts under the fan shroud where it is serviced by a three-prong connector plug. Open the plug and connect one test lead from an ohmmeter to the red lead pin and the other to the underside of the shroud. Observe the meter and reverse the test leads. The meter should report high resistance in one direction and no resistance in the other. Do the same for each black lead pin.

The lighting circuit should be tested with a number 4001 headlamp connected between the output terminal and a reliable engine ground. The lamp should burn brightly at medium engine speeds. If it does burn brightly, the problem is in the external circuit between the engine and the vehicle lights. If the lamp does not light or burns feebly, the problem is in the alternator. Check coil continuity with an ohmmeter as shown in Fig. 6-13. High or infinite resistance means a defective stator.

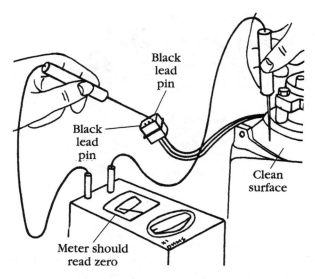

FIG. 6-11. *Testing for a shorted charging coil on the dual-circuit alternator.* Briggs & Stratton Corp.

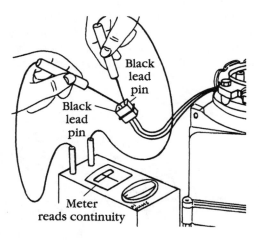

FIG. 6-12. *Testing charging-coil continuity on the dual-circuit alternator.* Briggs & Stratton Corp.

The 10 A alternator

Used on the series 200400 and 320400 engines, the 10 A alternator is a heavy-duty device delivering better than 4 A at 2000 rpm and a full rating at

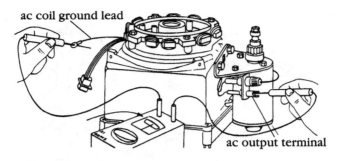

FIG. 6-13. *Testing ac circuit continuity on the dual-circuit alternator.* Briggs & Stratton Corp.

3600 rpm. The regulator is more flexible than those used on the smaller engines and can handle large loads without overcharging the battery.

Check voltage across the battery terminals with the engine turning at full-governed rpm. Less than 14 V on a fully charged battery means stator or regulator rectifier problems.

Disconnect the plug at the regulator rectifier and connect an ac voltmeter to each of the two outside plug terminals (Fig. 6-14). A reading of less than 20 V per terminal means a defective stator. Check the regulator rectifier by default. That is, if the system fails to deliver sufficient charging voltage and the stator appears okay, replace the regulator rectifier.

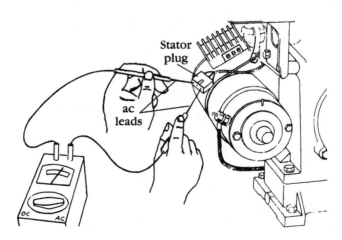

FIG. 6-14. *Testing the stator on the alternator.* Briggs & Stratton Corp.

The Nicad system

An option on 92000 and 110900 engines, the Nicad system consists of a gear-driven starter motor, starter-ignition switch, plug-in battery charger, and a 12 V nickel-cadmium battery. Nicad systems are intended for rotary lawnmower applications so the starter-ignition switch is mounted on the handlebar where it is electrically isolated from the engine. The switch stops the engine by grounding the magneto primary circuit through the connector clipped on the engine shroud. If the connector comes free of its clip, the magneto will be denied ground but the engine will continue to run regardless of the switch setting.

The first place to check if trouble arises on these engines is the battery. Nickel-cadmium cells are by no means immortal. Load the battery with two G.E. number 4001 sealed-beam headlamps connected in parallel (Fig. 6-15).

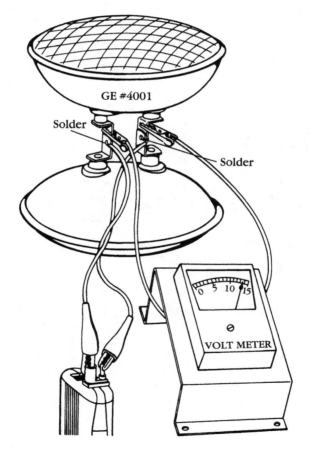

FIG. 6-15. *Testing the Nicad battery.* Briggs & Stratton Corp.

Monitor the voltage. The meter should show at least 13.6 V after one minute of draw. Readings of 13 V and less mean that one or more of the cells are defective. The lights should burn brightly for at least five minutes.

The half-wave rectifier supplied with this system should be capable of recharging a fully depleted battery over a period of 16 hours. An inexpensive tester can be constructed from the following materials:

- 1 1N4005 diode
- 1 red lamp socket (Dialco No. 0931-102)
- 1 green lamp socket (Dialco No. 0932-102)
- 1 neon bulb, No. 53
- 1¾-in. machine screw, No. 6-32
- 1¾-in. machine screw, No. 3-48

Wire the components as shown in Fig. 6-16. If neither bulb lights, the transformer or charger diode is open. If both bulbs light, the charger diode is open and passing alternating current. A properly working charger will light only the green bulb.

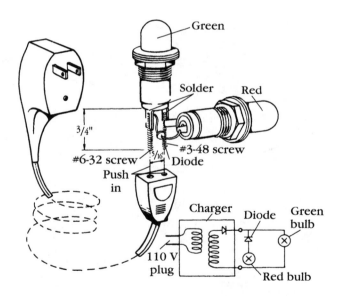

FIG. 6-16. *A homemade rectifier tester.* Briggs & Stratton Corp.

7

Engine mechanics

This chapter describes repair and overhaul procedures for all single-cylinder Briggs engines, except Japanese-built Vanguard models and the just-released Micro edger engine, for which information is lacking.

Major engine repairs involve specialized tools, some of them available only from the factory. Others can be purchased from industrial tool houses or fabricated at home. These tools are described as the need for them arises.

Cleanliness is, of course, paramount. A self-service car wash makes a pretty good degreasing facility for engine blocks and other large castings. Biodegradable cleaners, such as Greased Lightening (around $13 a gallon) or the more expensive Loctite Natural Blue, are safe for the surface we care about most, easy to dispose of, and eliminate all risk of fire. You will also need a supply of lintless rags and a clean, flat area to lay out parts in order of disassembly.

Some bolts and screws penetrate into the crankcase. To avoid oil leaks, the threads should be coated with a sealant, such as Fel-Pro Plia-Seal or General Motors PN 1050026. Old-timers swear by Permatex No. 2 and, truth be told, it works as well as any.

Vertical-shaft engines are awkward to handle while the crankshaft is in place. The fixture shown in Fig.7-1 provides at stable platform clear of bench clutter.

Resources

Buying parts on the Internet can save a few dollars, but robs your local Briggs dealer of incentive to provide assistance and advice. Engine work, like other trades, is characterized by an informal network of information exchange. Being on friendly terms with the dealer helps.

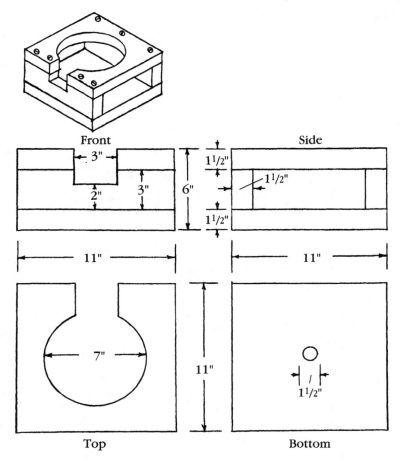

FIG. 7-1. *A work stand for vertical-shaft engines can be constructed from scrap lumber.* Briggs & Stratton Corp.

But dealers are not the last word when it comes to machine work or precision tuning. Automotive machinists have access to more sophisticated tooling and wider expertise than dealer mechanics, who rarely go beyond factory fixes.

Material on the factory web site www.briggsandstratton.com can also be very helpful. Briggs provides exploded views of all recent-production engines, together with parts numbers and suggested list prices. Briggs can also supply short blocks, i.e., new blocks assembled and all internal parts, less heads, flywheel, and accessories, for most engines. Short blocks come with a factory warranty.

Diagnosis

Begin by trying to assess the amount of damage. Major areas of concern are:

- Black, carbonized crankcase oil: Expect to find severe and often non-repairable damage on all bearing surfaces. Unless you have an ample stock of used parts, a heat-discolored conn rod is evidence enough to scrap the engine.
- Black, but still-liquid crankcase oil: Because of its location, the lower main bearing on vertical-shaft engines will exhibit most wear, followed by the connecting rod and rod journal.
- Traces of grit and sand in the carburetor throat: A leaking air filter or filter gasket results in rapid cylinder wear, especially on aluminum-bore engines. At the minimum, you will need to hone the cylinder clean of imbedded grit and replace piston rings.
- Loss of power: A blown-head gasket is the usual culprit, although any failure of fit, sealing, or lubrication costs power. See Fig. 7-2.
- High oil consumption: Figure 7-3 covers all bases, but it is important to distinguish between loss of oil caused by burning and oil leaks. Blue exhaust smoke suggests ring wear. A leaking intake-valve oil seal on OHV models smokes the exhaust immediately after startup and, often, during sudden acceleration. Failure of the breather (see below) also results in high oil consumption. Oil leaks can often be repaired without major disassembly.
- Bent crankshaft: Visible wobble and the accompanying vibration mean the crankshaft is bent. The only safe cure is to replace the part.

Warning: the factory strongly opposes the practice of straightening bent crankshafts because of the risk of subsequent failure. If the crankshaft in question is attached to a rotary lawnmower blade, the results could be tragic. (See the Crankshaft section for further discussion.)

Manometer

A pair of capscrews secure the crankcase breather to the block in the area above the camshaft. The breather draws a slight vacuum on the crankcase and, on modern engines, recycles combustion byproducts into the carburetor. The latter function explains why a poorly maintained engine can sometimes run when the gas tank is empty. The oil contains enough raw fuel to support combustion.

The breather assembly includes baffle and a wire-mesh element that function as a vapor separator to limit the amount of oil vented, and a check valve in the form of a fiber disc. As the piston moves toward bottom dead center (bdc), pressure in the crankcase rises and the check valve opens. The

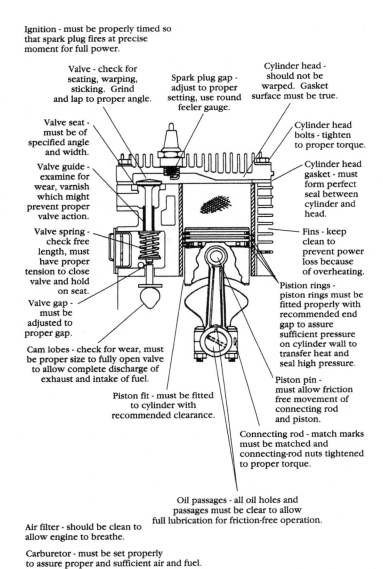

Ignition - must be properly timed so that spark plug fires at precise moment for full power.

Valve - check for seating, warping, sticking. Grind and lap to proper angle.

Spark plug gap - adjust to proper setting, use round feeler gauge.

Cylinder head - should not be warped. Gasket surface must be true.

Valve seat - must be of specified angle and width.

Valve guide - examine for wear, varnish which might prevent proper valve action.

Valve spring - check free length, must have proper tension to close valve and hold on seat.

Valve gap - must be adjusted to proper gap.

Cam lobes - check for wear, must be proper size to fully open valve to allow complete discharge of exhaust and intake of fuel.

Cylinder head bolts - tighten to proper torque.

Cylinder head gasket - must form perfect seal between cylinder and head.

Fins - keep clean to prevent power loss because of overheating.

Pistion rings - piston rings must be fitted properly with recommended end gap to assure sufficient pressure on cylinder wall to transfer heat and seal high pressure.

Piston pin - must allow friction free movement of connecting rod and piston.

Piston fit - must be fitted to cylinder with recommended clearance.

Connecting rod - match marks must be matched and connecting-rod nuts tightened to proper torque.

Oil passages - all oil holes and passages must be clear to allow full lubrication for friction-free operation.

Air filter - should be clean to allow engine to breathe.

Carburetor - must be set properly to assure proper and sufficient air and fuel.

FIG. 7-2. *Factors that affect power output.* Tecumseh Products Co.

valve closes when pressure falls on the upstroke. The resulting vacuum helps control oil consumption.

While professional mechanics test for crankcase vacuum after the fact, when a just-rebuilt engine smokes like a mosquito fogger, a beginner might

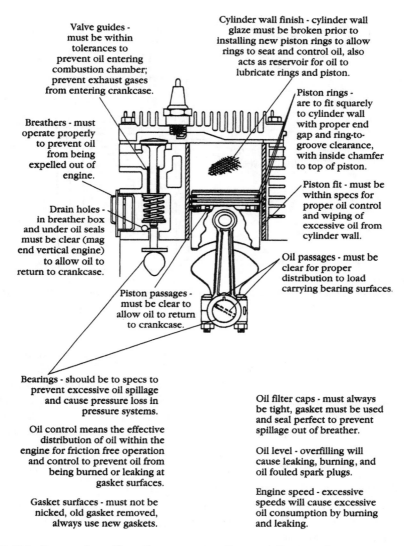

Valve guides - must be within tolerances to prevent oil entering combustion chamber; prevent exhaust gases from entering crankcase.

Cylinder wall finish - cylinder wall glaze must be broken prior to installing new piston rings to allow rings to seat and control oil, also acts as reservoir for oil to lubricate rings and piston.

Breathers - must operate properly to prevent oil from being expelled out of engine.

Piston rings - are to fit squarely to cylinder wall with proper end gap and ring-to-groove clearance, with inside chamfer to top of piston.

Drain holes - in breather box and under oil seals must be clear (mag end vertical engine) to allow oil to return to crankcase.

Piston fit - must be within specs for proper oil control and wiping of excessive oil from cylinder wall.

Oil passages - must be clear for proper distribution to load carrying bearing surfaces.

Piston passages - must be clear to allow oil to return to crankcase.

Bearings - should be to specs to prevent excessive oil spillage and cause pressure loss in pressure systems.

Oil control means the effective distribution of oil within the engine for friction free operation and control to prevent oil from being burned or leaking at gasket surfaces.

Gasket surfaces - must not be nicked, old gasket removed, always use new gaskets.

Oil filter caps - must always be tight, gasket must be used and seal perfect to prevent spillage out of breather.

Oil level - overfilling will cause leaking, burning, and oil fouled spark plugs.

Engine speed - excessive speeds will cause excessive oil consumption by burning and leaking.

FIG. 7-3. *Factors that affect oil consumption.* Tecumseh Products Co.

want to make this test first. You can use a vacuum gauge, but the home-made manometer shown in Fig. 7-4 is more precise.

The unit consists of a transparent plastic tube, an adapter for the oil-filler boss and a shutoff clamp (not shown) mounted on a board with a scale marked off in inches. A hose adapter can be made from an oil-filler plug and a piece of copper tubing, or from a rubber or cork stopper for engines

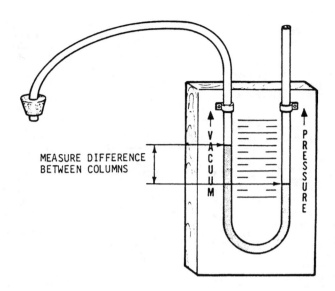

MEASURE DIFFERENCE
BETWEEN COLUMNS

FIG. 7-4. *U-tube manometer. At near wide-open throttle, engine vacuum should be 5–10 in. of water.* Kohler.

with a dipstick. The working fluid is water colored with food dye for visibility.

Insert the stopper into the oil-filler boss, close the shut-off clamp, and run the engine at 3200 rpm or so. Open the clamp and observe the reading, which should be between five and ten in. higher on the vacuum side of the tube than on the side open to the atmosphere. Close the clamp before shutting down the engine.

Zero vacuum or positive pressure means:

- A clogged crankcase breather screen or vent line
- An inoperative breather check valve
- Leaks at the dipstick-tube o-ring, flange gasket, oil-gallery cover (OHV engines), or crankshaft seals
- Blowby caused by worn rings and/or cylinder bore and
- Excessive exhaust back pressure. This condition is rare, but can occur with the newer, low-noise mufflers.

Wash the breather assembly in solvent, let it drain and check valve clearance. When unseated, the valve should clear the housing by 0.045 in. Figure 7-5 illustrates how a spark-plug gauge is used to determine this clearance. Do not force the issue and risk distorting the valve retainer—if the disc unseats 0.040 in. or so, the breather is probably okay.

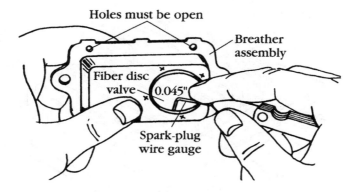

Holes must be open

Breather assembly

Fiber disc valve

0.045"

Spark-plug wire gauge

FIG. 7-5. *Use a round spark-plug feeler gauge to determine the amount of movement in the check valve.* Briggs & Stratton Corp.

Cylinder head

Note the bolt lengths. Aluminum-block side-valve engines typically have longer bolts in the exhaust-valve area. If the head sticks, break the gasket seal with a rubber mallet. Inspect the head gasket for breaks and the castings for carbon tracks that indicate leak paths. A blown-head gasket is usually an isolated phenomenon, but it can mean a warped head or pulled bolt threads.

Head-bolt and spark-plug threads can be renewed with threaded inserts. Because the tools are expensive ($43 for the Time-Sert 5/16-in. × 16 kit and $140 for HeliCoil M-14 × 1.25 mm spark-plug kit) thread restoration is best left to automotive machinists.

Carefully scrape off carbon deposits and gasket residue.

Warning: Until the early 1970s, Briggs engines came with asbestos head gaskets, and dealers continued to sell these gaskets for years afterward. Most were dull gray in appearance, but short of microscopic examination, it's impossible to determine the gasket composition. When in doubt, soak gasket with oil and carefully cut away the shards with a single-edged razor blade. Do not use a wire wheel or any other tool that would raise dust. Try to find a safe manner of disposal.

Check the head for distortion as shown in Fig. 7-6. Using a piece of plate or mirror glass as a work table, attempt to insert at 0.002-in. feeler gauge between bolt holes. If the gauge passes, the head should be resurfaced. The least expensive way to generate a flat surface is an "Armstrong mill," which consists of a sheet of # 360 wet-or-dry abrasive paper taped to the glass. Move the head over the abrasive in a figure-8 pattern, while pressing down

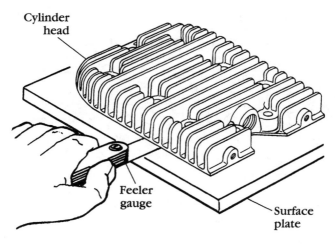

Cylinder
head

Feeler
gauge

Surface
plate

FIG. 7-6. *Over-tightening head bolts can bow the casting and invite leaks. Resurface if a 0.002-in. feeler blade can be inserted past the gasket flange and surface plate.* Kohler.

on the center of the casting. The same technique can be used for carburetor flanges and other distortion-prone parts.The new head gasket goes on dry, as do all Briggs & Stratton gaskets.

Clean bolt threads and lubricate prior to assembly. If you are a perfectionist, chase the block threads with a good quality (and accurate) bottoming tap. Although these engines do not develop high levels of combustion pressure, use of the correct lubricant on the bolt threads is necessary to obtain reasonably correct torque readings. At one point the factory recommended graphite grease or PN 93963, a valve-guide lubricant, for this application. They have since dropped the graphite-grease option, but it appears that either could still be used. In any event, apply the lubricant sparingly (to prevent air pockets from forming) to the threads and to the undersides of the bolt heads.

Figure 7-7 shows the tightening sequences for the various flatheads. OHV heads for engines under discussion have either four or five bolts. Tighten in a diagonal fashion as if you were mounting a wheel on a car. Run down the bolts in three increments: one-third torque, two-thirds torque, and full torque as specified in Table 7-1.

Valves

Figures 7-8 and 7-9 illustrate side and overhead valves. Umbrella-type oil seals are used on the intake valves for all OHV engines and on the exhaust valves of some Vanguards. Puffs of blue smoke immediately upon startup

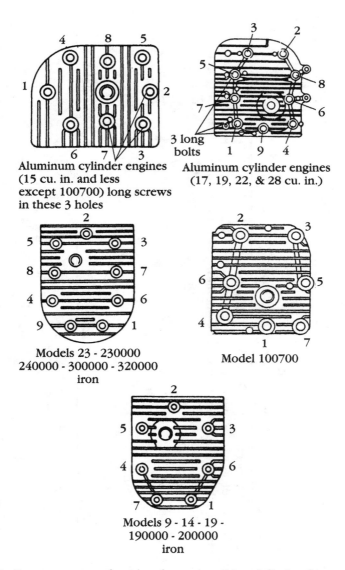

Aluminum cylinder engines
(15 cu. in. and less
except 100700) long screws
in these 3 holes

Aluminum cylinder engines
(17, 19, 22, & 28 cu. in.)

3 long
bolts

Models 23 - 230000
240000 - 300000 - 320000
iron

Model 100700

Models 9 - 14 - 19 -
190000 - 200000
iron

FIG. 7-7. *Torque sequences for side-valve engines.* Briggs & Stratton Corp.

mean that oil had been drawn past the seal when the engine was last shut down.

If either valve leaks, the engine may not start. If it does start, power will be down, although no-load rpm may be unaffected. A leaking intake valve can "pop-back" through the carburetor, mimicking the effect of a lean mixture.

Table 7-1
Side-valve engine torque limits

Engine block material	Cubic-inch displacement (1st one or two digits of model number)	Cylinder-head bolts (lb/in.)	Flange or crankcase cover (lb/in.)	Conn rod bolts (lb/in.)	Flywheel nut (lb/ft) 1 lb/ft = 12 lb/in.
Aluminum	6, 8 9, 10 (except as noted below), 12	140	85	100	55 lb/ft
	100200, 100900, 13	140	120	100	65 lb/ft
	14	165	140	165	65 lb/ft
	17	165	140	165	65 lb/ft
	19, 22, 25	165	140	185	65 lb/ft
	28	165	185	185	90 lb/ft
Cast iron	23, 24, 30, 32	190	n/a	190	135 lb/ft

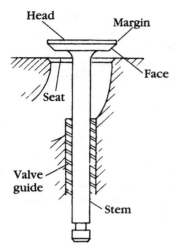

FIG. 7-8. *Side-valve nomenclature.* Clinton Engines Corp.

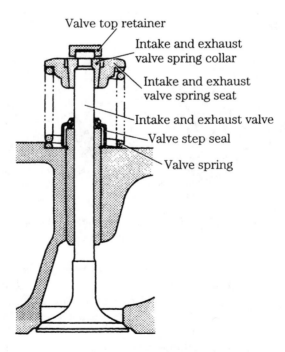

FIG. 7-9. *All Briggs OHV models have an umbrella-type oil seal on the intake valve, and several use the same type of seal on the exhaust.* Yanmar.

Carbon collects under the intake-valve head, especially on lightly loaded engines. Rich mixtures can leave gum, or varnish, deposits on the valves. Varnish deposits severe enough to stick the valve in its guide nearly always result from using stale gasoline. If a valve sticks on a side-valve engine, the cam forces the valve open, where it acts as a massive compression leak. On an OHV engine, the valve usually remains seated, and the associated pushrod bends. The engine will have good compression, but no fuel entry.

As the hottest spot in the engine, the exhaust valve is prone to corrosion and leakage across the seat. Water in the fuel contributes to the corrosion problem. A dull black, polished appearance means the valve has failed to seat properly and, consequently, has overheated. High exhaust-valve temperatures also result from lean mixtures and clogged cooling fins in the valve area. Brown, tan, or yellowish under-head deposits are normal, their color determined by additives in the fuel.

Valve removal

Most modern engines retain their valves with split collets similar to those used for automobile valves. Older examples use a collar with an elongated slot (Fig. 7-10) and, going back even further, a pin that fits into a cross-drilled hole in the valve stem.

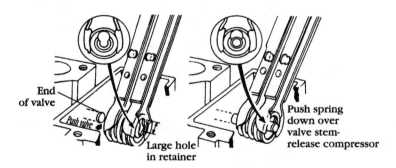

FIG. 7-10. *Installing Briggs one-piece valve retainers requires the use of a valve-spring compressor.*

Intec and other American-made OHV engines employ stamped-steel rocker arms mounted on pedestals and secured by nuts that double as fulcrums. These nuts control the amount of valve lash and are locked into place by Allen or Torx screws. Loosen the screws to release pressure on the threads before loosening the fulcrum nuts.

Once the nuts are removed, the rocker arms and pushrods can be lifted free. Note that pushrods fit into sockets in the valve lifters that are buried deeply within the block. Make sure the pushrods seat upon assembly.

The drawing referenced in Fig.7-10 illustrates the Briggs & Stratton PN 19063, probably the finest valve-spring compressor available for small side-valve engines. Without this tool, you can make do with a pair of screwdrivers, but the work is awkward. Figure 7-11 illustrates how split keepers on OHV engines are released. Support the valve head with a soft wooden block.

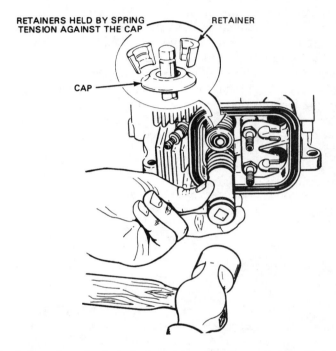

FIG. 7-11. *Release the valve locks by impacting the valve cap with a $^3/_4$-in. deep-well socket.*

Do not mix intake- and exhaust-valve components. Wire brush the carbon deposits off and, if gum is present on the stems, soak the valves in carburetor cleaner. Using a pipe cleaner, swab down the valve guides with the same toxic stuff.

Warning: Carburetor cleaner, either in the aerosol or liquid form, is not good for living things. Protect your eyes and hands, and work in a well-ventilated place.

The free-standing height of a used spring should be at least 90% of the height of a new spring. The spring should stand upright; tilt means that the coils have weakened on one side. Surface pitting is an early indicator of fatigue.

Some valve-related components are difficult or impossible to obtain. In an emergency, one can, for example, straighten bent pushrods and have worn guides rebushed.

Valve installation

On ohv models, place the head on a wood block to support the valves while the springs and keepers are installed. If PN 93963 valve-guide lubricant is used, it should be confined to the valve stems. Otherwise, assemble the parts with motor oil. Position the valve in the head, slip the the seal over the stem and valve-guide boss, and install the valve spring and the valve-spring retainer washer. Press down on the spring, collapsing it enough to install the two keepers. Figure 7-12 illustrates an easily fabricated a valve-spring compressor of the type needed for the stiff valve springs used on several large Intec engines. Thumb pressure suffices for the others. Release the spring and verify that the both keepers have seated in their grooves. A keeper that stands proud has failed to seat.

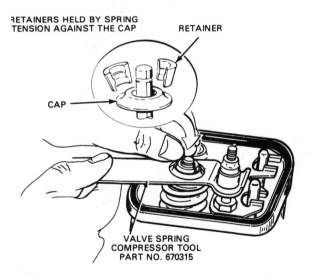

FIG. 7-12. *This drawing provides a good view of collet-type split keepers and the tool used for installation.* Tecumseh Products Co.

Install the pushrods, rocker arms, and fulcrum nuts. Adjust the lash as described in the section "Lash Adjustment."

On side-valve models, compress the valve spring with PN 19063 or the equivalent tool, and place the assembly in the valve chamber. Insert the valve and holding it down with your thumb, insert the keepers. On older engines, slip the collar home to lock the spring to the valve. The two-piece collet-type keepers used on later models can be a bit tricky to install in the confines of the valve chamber (Fig. 7-13). Automotive parts houses can supply a magnetic keeper-installation tool that makes things less frustrating. Once the keepers go home, release spring tension. Verify that the keepers seat in the valve stem groove.

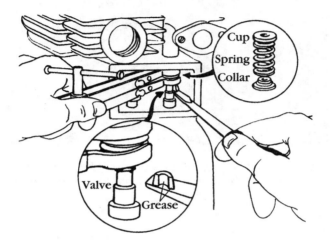

FIG. 7-13. *Installing split collets with the aid of grease.* Briggs & Stratton Corp.

Lash adjustment

The fulcrum nut sets the lash on overhead valves (Fig. 7-14). Failure to back off the Allen or Torx screw in the head of the nut will result in a sheared nut and bad language. Remove the spark plug and using a screwdriver as a probe, turn the flywheel to bring the piston about a quarter-inch past top dead center (tdc) on the compression stroke. At this point, both valves should be closed.

Valve lash for single-cylinder OHV engines (except Vanguard)	
Intake	**Exhaust**
0.005 in.	0.006 in.

To measure the lash on side-valve engines, turn the flywheel to bring the piston a fraction of an inch past tdc on the compression stroke, as described earlier. Insert a feeler gauge between the valve stem and tappet (Fig. 7-15).

Side-valve lash is determined by the location of the valve seat relative to the tip of the valve stem. A new valve or a reground valve or seat reduces the lash. To compensate, grind the stem flat, "kissing" the wheel just hard

FIG. 7-14. *Valve last on OHV engines is the clearance between the rocker arm and the valve stem. Tightening the fulcrum nut set screw fixes the adjustment.*

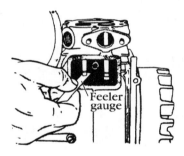

FIG. 7-15. *Valve lash on a side-valve engine is the clearance between the tappet and valve stem.* Clinton Engines Corp.

enough to remove a few thousandths. It is disconcertingly easy to remove too much metal. Check the work frequently. Finish with a hand stone, breaking the edges to remove the burr left by the wheel.

Excessive lash can only be corrected with a new valve or by regrinding the valve face and/or seat.

Valve lash for side-valve engines		
	Intake	**Exhaust**
Aluminum block		
6–10 cid	0.006 in.	0.008 in.
11 cid and larger	0.006 in.	0.010 in.
Cast-iron block		
23 cid and larger	0.008 in.	0.018 in.

Guides

Valve guides tend to bell-mouth, a condition that makes valve sealing problematic. The factory supplies an array of go-nogo gauges for determining guide wear, but if the valve exhibits perceptible wobble in the full-open position, the guide is worn. A 64th of an inch is enough to make one consider replacement. Regardless of condition, replace the guides before doing any machine work on the valves or seats.

Many Briggs engines do not have replaceable guides, as such. Instead the guide is reamed oversize and a bushing installed. Other guides are driven out with a punch and replaced as a unit. In any event, the bushing or guide ID must be finish-reamed to size with a factory tool that cannot be purchased elsewhere. The reamer leaves the guide a few thousandths of an inch larger than the valve OD, which on modern engines is often metric. Most amateur mechanics should farm this and all other valve work to an experienced auto machinist or to a Briggs dealer. As with all machine work, have the parts in hand before you begin. If parts are not available, bronze replacement bushings can be fabricated.

Valve grinding

Valve grinding involves the use of a valve lathe and a high-speed seat grinder. While the machinist will have final say on seat angles, it appears that Briggs uses a 45° angle on both the valve face and seat, without the 1/2° of interference angle that has become standard for automotive work. The valve margin should be at least 1/64-in. wide after grinding.

Valve seats

Cracked, severely worn or loose valve seats must be replaced. Most mechanics extract faulty seals with a puller, such as the one shown in Fig. 7-16. This might explain why seats sometimes leak after installation. A better approach is to have your machinist cut the seat out. The machinist will press in a new seat and, on aluminum blocks, peen it for security (Fig. 7-17).

Flange

The flange, or crankcase cover, supports the pto ends of the crank and camshaft, and, on vertical-shaft engines, locates the oil slinger. It also houses the oil pump on OHV engines.

Disassembly

Remove the pulley or blade adapter, together with locating keys and other bits of hardware that might be present on the end of the crankshaft. Blade

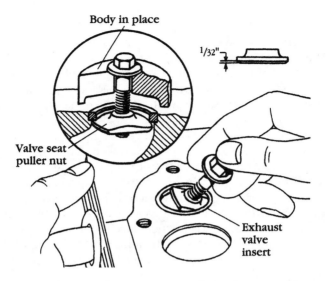

FIG. 7-16. *Most mechanics extract worn or cracked valve seats with a puller.* Briggs & Stratton Corp.

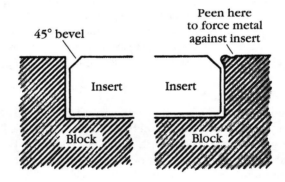

FIG. 7-17. *Peening the valve seat helps to secure it on aluminum blocks. Even so, it is by no means unknown for a valve seat to leak around its OD when the block heats and expands.* Briggs & Stratton Corp.

adapters can be stubborn and removal might require use of a gear puller and liberal quantities of penetrating oil. A bolt in the center hole protects the crankshaft threads from contact with the puller. Occasionally an adapter will defeat even the strongest puller, and must be hammered off. Using two heavy hammers, direct simultaneous blows at the adapter neck, driving it down an off.

Remove the rust from the crankshaft stub with strip sandpaper. Dress keyways with a file to dull the sharp edges that might damage the oil seal. In addition, the extreme end of a hard-used crankshaft sometimes bulges under center-bolt torque. A few thousandths of an inch is enough to score the flange bearing during disassembly. Crank bulging calls for a sharp file, strip sandpaper, and patience.

Rotary lawnmowers bend crankshafts. The troubleshooting chapter describes a quick and dirty way to detect bends; a dial indicator, positioned near the end of the crank, will provide a precise reading. Although few used crankshafts can meet the factory specification of ± 0.001 in., the greater the dislocation, the more the vibration.

Remove the flange hold-down bolts. Make a sketch if bolt length varies or if certain bolts have sealant on their threads. Some engines have numbered bolts to encourage the mechanic to follow the correct torque sequence upon assembly. Be certain you know where No. 1 bolt goes.

If you are unfamiliar with the engine, position it on the bench with the flywheel down. This should keep the cam and other internal parts in their proper places as the flange comes off.

Lightly oil the crankshaft stub and, using a soft mallet, gently tap the flange to break the gasket seal. Once the locating dowels disengage, the flange should slide easily over the shaft. If the flange binds, stop and polish out the high spot on the crankshaft.

As a matter of good practice, turn the flywheel to align the timing marks before further disassembly. If there is any ambiguity, clarify the situation before proceeding. Refer to the "Camshaft" section.

Clean the flange and inspect the lower main bearing, which tends to fail earlier than the magneto-side bearing. Plain bearings (i.e., DU™ bronze, steel-backed aluminum, or the flange itself) should be smooth and without deep scores or discoloration. Most mechanics judge wear by the fit on the crankshaft. If a ball bearing is used, clean it thoroughly to remove all traces of lubricant and spin the inner race by hand. It should turn easily with no rough or hard spots.

Assembly

Verify timing-mark alignment, slip the oil slinger over the camshaft stub (with the spring washer if used) and mount a new flange gasket on the block. RTV sealant goes on bolts that originally used it 30W motor oil on the others. On OHV engines, align the slot in the oil-pump shaft with the camshaft drive pin. Gently tap the flange over the indexing dowels. Do not force matters: when everything is in alignment, the flange goes home with an audible click. Tighten the fasteners in multiple stages, working in a star pattern. Torque sequences are not included here because, like the fabled talking cat, nobody listens.[1] Use the gap at the gasket interface as a guide to draw down the fragile casting down squarely and evenly. Table 7-1 lists torque limits.

At this point, you may want to check crankshaft end play. The minimum permissible amount of float is 0.002 in. to allow for thermal expansion. The maximum varies with model and application. While 0.008 in. is the safe maximum for all engines, including direct-drive pump applications, some (e.g., 9-cid lawnmower engines) tolerate more than 0.030 in. Consult your dealer.

Measuring movement along the crankshaft's axis requires dial indicator or a sheave and a feeler gauge (Fig. 7-18). Bump the crankshaft to its travel limit one direction, zero the dial indicator, or add leaves to the feeler gauge for zero clearance, and bump the crank in the opposite direction.

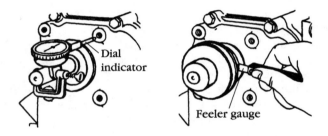

FIG. 7-18. *Crankshaft end play varies with the model, but 0.008 in. is safe for all engines.* Briggs & Stratton Corp.

The thickness of the flange gasket determines how far the crankshaft thrust face stands off from the flange bearing surface. All models require a 0.015-in. thick flange gasket, except 230000 and 240000 side-valve engines that use a 0.020-in. gasket. These standard gaskets must be in place to prevent oil and air leaks. A second standard gasket or some combination of 0.005- and 0.009-in. gaskets can be added to increase the end play to the 0.002-in. minimum specification (Fig. 7-19).

Excessive end play results from normal wear and the thrust loads generated by rotary lawnmower blades. A hardened steel thrust washer acts as a shim to limit crankshaft motion. On plain-bearing engines, the washer goes

---

[1] According to the 19th century story, there was a fellow who believed his cat knew how to talk, but was too stubborn to do so. So he bribed the cat with caviar, lobster, and other delicacies, spending all of his money and going hungry himself. One day the cat looked up and said, "The roof's going to fall in." The cat walked out, but the man stood there transfixed with joy. "He talked! My cat actually talked.!" At that point the roof collapsed. The cat looked back at the wreckage and said, "Huh! Why should I talk when nobody listens to me?"

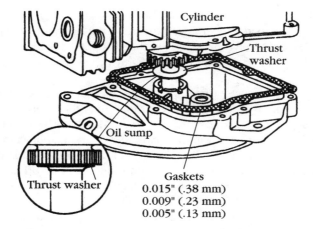

FIG. 7-19. *Thrust washers reduce crankshaft end play and some combination of stacked flange gaskets increases it.* Briggs & Stratton Corp.

between the crankshaft thrust face and the flange. If the pto end of the crank rides on a ball bearing, the thrust washer goes on the plain-bearing magneto side. The bearings must be replaced to restore end play for engines that support the crankshaft on two ball-bearings.

Oil seal

Leaking flange or crankcase cover seals are pried loose with a large screwdriver as indicated in Fig. 7-20. Note how deeply the seal is inserted into the

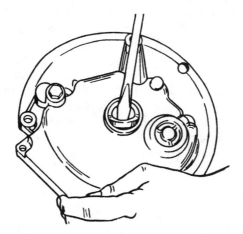

FIG. 7-20. *PTO and magneto-side oil seals come free with the help of a large screwdriver.* Tecumseh Products Co.

casting. Most are flush with the top of the boss; seals for ball-bearing engines go a fraction of an inch deeper. Insert the tip of the screwdriver deep into the elastomer element to give purchase on the U-shaped steel retainer. A sharp blow with the palm of the hand on the screwdriver handle will lift the seal out of its boss.

Examine the replacement seal. In profile, the elastomer element has the shape of a triangle, with one side steeper than the other. This is the pressure side and must be inboard, toward the crankcase. Most seals have their outboard, or installation, side inscribed with the part number and manufacturer's name.

Inspect the crankshaft. If the seal-contact area is scored, it is possible to offset the new seal a few hundredths of an inch from its original position, in order to provide an unblemished contact surface. But the repositioned seal must not block oil-return port on the casting.

Most new seals have a flexible coating on the OD to prevent leaks. Others do not. These seals require a very thin coating of Permatex No. 2 on their outer rims. (The viscosity of RTV makes it a poor choice for this application.) Dab a bit of sealant on your finger and wet the OD, being careful not to contaminate the element or leave a surplus that could block the oil port.

A wood block, large enough to cover the seal diameter, makes a good installation tool for flush-mounted seals. Those that are countersunk require a driver, accurately sized to the diameter of the boss. A smaller tool distorts the seal and results in early failure. Once you have the proper tool, drive the seal home with light hammer taps.

Pack the elastomer element with assembly grease to provide lubrication during initial startup. A single layer of Scotch tape over the crankshaft key-ways stands in for the Briggs seal protector tool. Keyway edges are sharp and, unless covered, can damage seal lips. Install the flange or crankcase cover, tighten as described earlier, and remove the tape.

It is also possible to replace the seal without disassembling the engine. Polish the crankshaft stub to remove all traces of rust and, using a small, dull chisel, collapse the outboard face of the seal. Work around the periphery, driving the metal face inward. Be very careful not to score the bore. Eventually the seal will relax its hold and come free.

Replace the seal as described above. A hardwood block, drilled to pass over the crankshaft end, serves as a driver.

Oil slinger and pump

Splash, generated by a camshaft-driven slinger on a vertical-shaft engines or a dipper on horizontal models, provides lubrication to engine internals. Figure 7-21 illustrates a typical slinger which, in this particular application, incorporates a spring washer between its bracket and the flange.

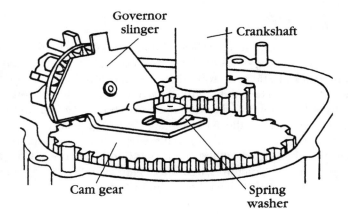

FIG. 7-21. *Models 100900 and 130900 use a spring washer between the bracket and flange.* Briggs & Stratton Corp.

Overhead-valve engines utilize a camshaft-driven pump to deliver oil to the camshaft gear and, depending upon engine model to camshaft and one or both main bearings. In some examples, the pump merely provides pressure to an oil filter. In the absence of an oil-pressure warning lamp, the assumption is that the pump works if oil reaches the valves. Figure 7-22 illustrates the workings of the Gerotor pumps used by Briggs & Stratton. Like the Wankel engine, these pumps are a marvel of counter-intuitive geometry.

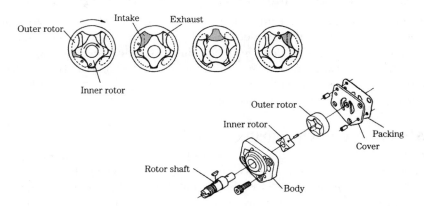

FIG. 7-22. *A Gerotor pump employs an eccentric inner rotor and a freely turning outer rotor to create a working chamber of variable capacity.* Yanmar.

Camshaft

This section appears more complicated that the work actually is. Briggs & Stratton uses three basic styles of camshafts, a one-piece iron version, a one-piece composite plastic and iron shaft, and an iron shaft that rides on a removable steel pin. The latter complicates timing a bit. In addition, all modern camshafts have some form of compression release acting on the intake and/or exhaust valve. But the average reader, with the average 6, 9, or 10-cid plain-bearing engine, need only to remember to index the timing mark on the cam gear with the mark on the crank gear during assembly. Compression releases come into the picture when the engine refuses to start and after all the more obvious suspects have been eliminated.

Aluminum-block, plain-bearing, side-valve and OHV engines

To disassemble, remove the flange, align the timing marks on the cam and crankshaft gears, and withdraw the camshaft. Use a Magic Marker to identify the intake and exhaust tappets, or valve lifters, for assembly. Many late-production engines employ composite plastic and iron cams. Plastic lobes explain why OHV-engine valve springs are weak enough to compress with a finger.

Clean the parts and blow out drilled oil passages in OHV cams. Lubricate cam lobes, bearing surfaces, and tappets with motor oil. To assemble, insert the tappets in their proper bores and slip the camshaft home (Fig. 7-23). Verify that timing marks align. Some crankshafts have removable gears that, of course, must be installed with the timing mark visible.

Aluminum-block, ball-bearing, side-valve engines

These engines align the cam with a mark on the crankshaft counterweight (Fig. 7-24). Remove the connecting-rod cap and align the timing marks. Push the rod and piston deep into the cylinder for clearance, and withdraw the crank and camshaft together, with timing marks aligned. To assemble, lubricate all parts with motor oil, install the tappets, align timing marks, and insert the crank and camshaft as a unit.

Aluminum-block camshaft bearings

Occasionally the bosses in the block and flange that support the camshaft develop scores. The factory would have you replace these castings, which means a short block or a new engine. But everything is repairable, except perhaps a broken heart or the crack of dawn. Iron camshaft journals can be brazed, turned down on a lathe, and the castings reamed oversize to fit.

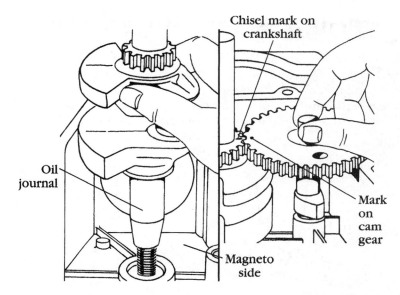

FIG. 7-23. *Briggs plain-bearing engines have timing marks stamped on the crankshaft-gear and camshaft-gear teeth.*

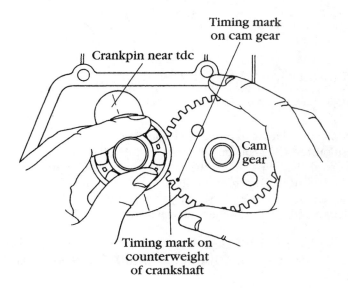

FIG. 7-24. *The camshaft on some ball-bearing engines is indexed to the crankshaft counterweight.* Briggs & Stratton Corp.

Cast-iron block, plain-bearing, horizontal-crankshaft engines

Remove the side plate and connecting-rod cap, and turn the crankshaft bring the counterweight clear of the camshaft gear. Withdraw the crankshaft, twisting and turning it as necessary.

The hollow camshaft rides on a steel pin that extends through both sides of the iron block. Drive the pin out from the pto side with a $^3/_8$-in. punch. Some resistance will be felt as the flywheel-side expansion plug is displaced (Fig. 7-25).

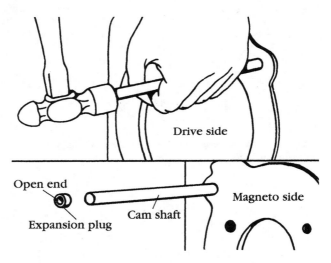

FIG. 7-25. *The cam on large-bore iron Briggs engines rides on a rod that passes through both sides of the block casting. Drive out the rod from the magneto side. Note that the expansion plug has its cupped side facing out.*

Assembly is the reverse process. Lubricate everything, install the tappets, and position the camshaft under them. If the valves remain in place, turn the cam so that it slips under the tappets without compressing the valve springs. The pin goes in from the flywheel side of the block. Do not force things: when aligned, the pin slides home easily. Coat the expansion plug with silicone and drive it home with a hardwood block.

Cast-iron block, ball-bearing, horizontal-crankshaft Models 9, 14, 19, 23, 190000, 200000, 230000, and 240000 engines

These engines are built on the same basic format as the cast-iron, plain-bearing models discussed earlier. The camshaft turns on a support pin, sealed by an expansion plug on the flywheel side of the crankcase. Disassembly procedures are similar, except the piston and rod must be extracted before model 240000 crankshafts are withdrawn.

Because the gears are on the flywheel side, the timing procedure varies from that described earlier. Remove the valves and color the crankshaft-gear teeth on either side of the timing mark with Prussian blue. Install the tappets with grease on their stems to hold them in place. Temporarily park the camshaft in the crankcase recess provided for this purpose (Fig. 7-26). Raise the cam to align the timing marks and, while holding the cam in mesh, install the supporting pin. Verify that the engine is in time.

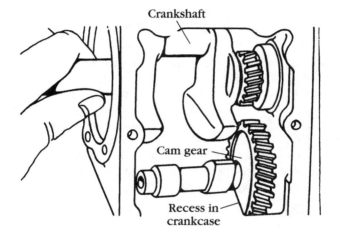

FIG. 7-26. *Many Briggs iron engines had a shelf-like recess in the block that supports the cam while the crankshaft is installed. Once the crank is in place, raise the cam, align the marks, and installed the camshaft rod.*

Coat the expansion plug with silicone and drive it home with a wooden block.

Cast-iron block, ball-bearing, horizontal-crankshaft — Models 300400 and 320400 engines

Remove the short bolt and spring washer from the pto-drive gear. Working from the magneto side of the block, loosen the long bolt two turns. Tap the head of the bolt to free the camshaft support rod. Continue to unscrew the bolt as the rod is forced out. When the pin is free, remove the two bolts that secure the diamond-shaped cam-bearing assembly (Fig. 7-27). While holding the camshaft with one hand so it doesn't fall, withdraw the bearing assembly. The camshaft is now free.

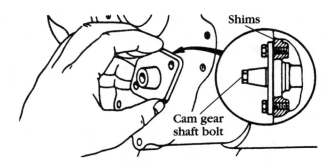

FIG. 7-27. *Shims fix camshaft end play on 30- and 32-cid iron engines.* Briggs & Stratton Corp.

Once parts have been cleaned, inspected for wear and damage, and liberally oiled, the assembly process can begin. Install the tappets and the breaker-point plunger in their respective bores. Insert the camshaft from the pto side and slide the support rod, blunt end first, through the pto bearing and into the camshaft. Mount the magneto-side bearing assembly and torque the two mounting screws to 85 lb/in. Thread the long bolt into the support rod and tighten it enough to hold the camshaft in place.

Check end play, which should be between 0.002 and 0.008 in. If there is insufficient end play (as can happen when a new cam is used), shim the bearing housing.

Shim thickness (in.)	Part number
0.005	270518
0.007	270517
0.009	270516

PN 299706 includes a new, slightly longer bearing and shims to correct excessive end play.

Once the crankshaft is in place, you will not be able to see the timing marks. Use Prussian blue to color the crankshaft gear tooth whose inboard edge indexes with the timing mark as shown in Fig. 7-28. Install the crank, aligning the referenced tooth with the mark on the camshaft gear. Main-bearing mounting screws are torqued to these specifications:

Magneto side—75-90 lb/in.
pto side—185 lb/in.

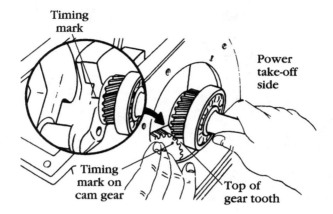

Timing mark

Power take-off side

Timing mark on cam gear

Top of gear tooth

FIG. 7-28. *Like some Kohlers, Briggs 30- and 32-cid iron engines employ disappearing timing marks that must be supplemented with a Magic Marker or Prussian blue on adjacent gear teeth.*

Automatic compression releases

Easy-Spin was the first and cleverest compression release and continues to be specified for the Classic and other "heritage" engines. However, emissions regulations make its future doubtful.

The device consists of a zero-lift zone on the cam lobe that prevents the intake valve from seating until late into the compression stroke. This represents a major loss of compression at low rpm, but as engine speed increases, the leak becomes less significant and power approaches normal. Blowback through the carburetor and its effect on mixture control appears to be the source the emissions problem.

Easy-Spin comes into play between 30° and 45° after bdc to prevent the intake valve from closing. The valve remains 0.015–0.017 in. off of its seat for a minimum of 30° of crankshaft rotation.

Overhead-valve engines (except Vanguard) and updated side-valve models use a centrifugal compression release that acts on the exhaust valve as shown in Fig. 7-29. The device consists of a spring-loaded yoke that cams open the valve at cranking speeds. Once the engine starts, centrifugal force retracts the yoke to restore normal valve action.

The manual starter engages the compression release used in conjunction with Easy-Spin on 11-cid, side-valve engines. Figure 7-30 illustrates the mechanism in the engaged position on the exhaust lobe.

Excessive valve lash defeats any of these compression releases to make starting difficult and, for larger engines, virtually impossible. Kickback on the starter cord and burned-out starter motors are secondary effects.

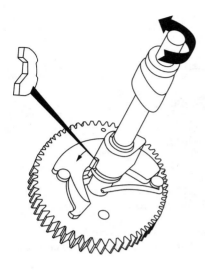

FIG. 7-29. *Briggs centrifugal compression release.*

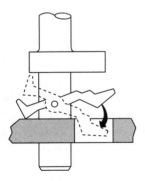

FIG. 7-30. *Another of the company's compression releases is engaged by the starter.*

Make any necessary adjustments to valve lash as described in the "Valves" section. Once the lash is correct, there is nothing more to go wrong with Easy-Spin. The other systems must overcome inertial and valve-spring forces in order to lift the exhaust valve. Cycle the mechanism by hand to establish that it engages and disengages smoothly, without binding. This should be enough to establish that starter-engaged compression releases function normally. Do the same for centrifugal releases. Examine the face of the ramp (the part that bears against the tappet) for wear.

Verify the action in an assembled engine by removing the valve cover. As the flywheel is turned (using the starter cord on 11-cid engines), observe the exhaust valve. A functional compression release unseats the valve a few thousandths of an inch late in the compression stroke and holds it open for 25° or so of crankshaft rotation. Wear on the compression-release ramp progresses to the point that valve lash intervenes to shelter it from opening forces. If this is the case, the ramp will lift the tappet or pushrod just far enough to close down lash without unseating the valve. The cure is a new camshaft.

Connecting rods

After many years of service, engines develop a step-like ridge on the upper end of the cylinder bore that "snags" the rings and must be removed before the piston can be extracted. A piston that will not be reused can sometimes be forced past the obstruction, saving the cost of purchasing a ridge reamer.

Turn the crankshaft to align the timing marks for reference upon assembly. Defeat the rod-bolt locks with a small punch, flattening them enough for wrench purchase (Fig. 7-31). Note the location of oil holes, the lay of the removable dippers used on some horizontal engines, and the way the match marks embossed on the cap and rod shank index (Fig. 7-32). These marks must align upon assembly to keep the rod journal round. Loosen the two conn-rod bolts in several increments and rotate the crankshaft a few degrees to disengage the cap. Lift off the cap and set it aside.

With the upper half of the connecting rod still riding on the crankpin, turn the crankshaft to bring the piston to tdc. Force the piston up and out with your fingers or a wooden dowel pin pressed against the underside of the piston. Replace the rod cap, lock and bolts, aligning the match marks.

Inspect the crankpin and big-end rod bearings for scratches, discoloration, and scores. A dull, satiny surface on the rod bearing is normal. Shop

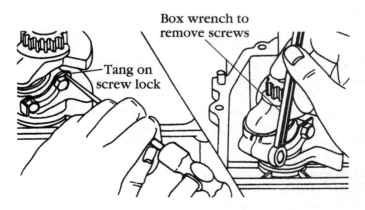

FIG. 7-31. *Disengage connecting-rod locks with a punch.* Briggs & Stratton Corp.

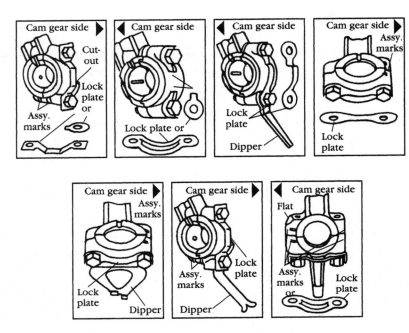

FIG. 7-32. *Briggs & Stratton uses several schemes to assure that rod shanks and caps are assembled correctly. But correctly assembled rods can be installed 180° out of phase. Should this happen, crankpin lubrication suffers and, on some models, the dipper collides with the block or camshaft.*

mechanics routinely accept light scratches and local bright spots that do not extend over more than 5% of the contact surface. Blue temper marks, deep gouges, or evidence of metal transfer mean that the rod was verging on seizure. Undersized rods available for newer models (and fit some earlier engines) permit the crankshaft to be reground. Otherwise, this kind of damage can only be rectified with a new crankshaft and rod.

Big-end bearing clearance is best determined with inside and outside micrometers, but inexpensive plastic gauge wire, available from car-parts stores, can substitute. Wipe the oil off the bearing surfaces and lay a piece of gauge wire along the length of the crankpin as shown in Fig. 7-33. Without turning the crankshaft, pull down the cap bolts in three increments to the specified torque.

Remove the cap and read the bearing clearance in thousandths by comparing the width of the wire with the scale printed on the package. The greater the clearance, the less the wire flattens.

Caution: plastic gauge wire has a shelf life of about six months before it hardens and becomes inaccurate.

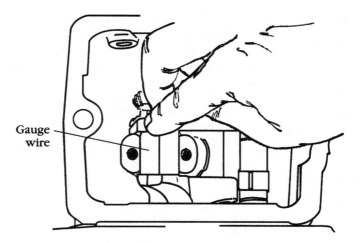

FIG. 7-33. *Plastic gauge wire placed for running clearance.*

The clearance should fall between 0.0015-in. for a new engine with tight assembly tolerances to 0.0030-in. for a middle-aged, but still serviceable example. Beyond 35 thousandths, one enters the domain of rod knock, which rapidly increases as the bearing pounds out.

Scrape off the remnants of the wire and repeat the operation, this time using two pieces of wire athwart the crankpin (Fig. 7-34). Differences in width between the two wires give a fairly accurate notion of taper.

Taper generates side loads that pass from the wrist pin to the piston. An engine intended for hard use should have zero or near-zero pin taper; one that

FIG. 7-34. *Plastic gauge wire placed for detection of taper.*

idles in semi-retirement can tolerate a bit more, but nothing approaching 0.001 in. As a rough rule of thumb, a crankpin/rod assembly with 0.0005-in. taper can pass muster if the piston skirt shows no unusual wear patterns.

A bent rod or a misaligned crankpin produces an hourglass-shaped wear pattern, 90° to the wrist pin (Fig. 7-35). A twisted rod rocks the piston leaving the wear signature shown in Fig. 7-36. Either of these wear patterns means that the rod and piston should be replaced and the crankshaft examined carefully. Pushing things a little further, you can determine the amount of bend by using the fire deck, or head-gasket surface, as a reference and a bar sized to fit the small-end rod bearing. Figure 7-37 illustrates the setup. Twist cannot be measured without an expensive jig.

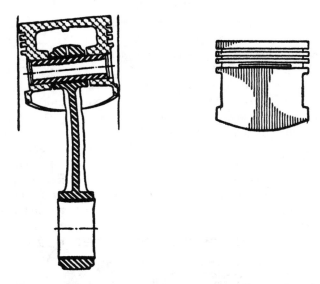

FIG. 7-35. *A bent connecting rod causes the piston oscillate and produce the wear pattern shown by the shaded areas.* Sealed Power Corp.

Wipe all traces of plastic gauge off the rod and crankpin, and flood the big- and small-end bearings with motor oil. A thin coat of oil on the piston skirt and rings suffices. Run the piston/rod assembly down into the bore as described in the following section. Install the cap, together with the associated hardware. Verify that:

- The piston/rod orientation is correct,
- Rod cap and shank match marks align and that
- The detachable dipper (when used) is installed correctly.

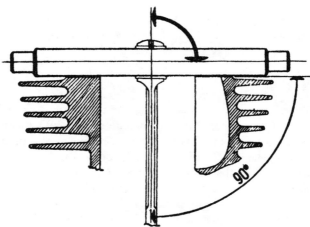

FIG. 7-36. *A twisted connecting rod causes the piston to rock and produces the wear patterns shown by the shaded areas.* Sealed Power Corp.

FIG. 7-37. *While few working mechanics have the leisure for this kind of inspection, a bent rod can be detected as shown here.*

Torque down the bolts evenly in at least three increments to the specification (Tables 7-1 and 7-2).

Turn the crankshaft through several complete revolutions. Binding suggests that the rod cap was reversed or that bearing surfaces are contaminated. Interference with the camshaft or other component means the rod

Table 7-2
Overhead-valve engine torque limits

Model name and cubic-inch displacement	Cylinder-head bolts (in./lb)	Flange or crankcase cover (in./lb)	Conn-rod bolts (in./lb)	Flywheel nut (lb/ft) 1 lb/ft = 12 lb/in.
Europa 9	160–65	85–90	100–105	60–65
Intec 11, 12	210–215	110–115	100–105	60–65
Intec 20	220–225	210–215	100–105	110–115
Intec 31	220–225	same as below	small bolt 125, large bolt 255	95–100
PowerBuilt 28*	220–225	bolts w/ lockwashers 150 bolts w/reduced diameter points 170	155	95–100

*No data on other PowerBuilt models.

was installed 180° out of phase. As a final check, attempt to slide the rod along the length of the crankpin. It should move under light finger pressure.

Piston

The thrust faces—the contact areas 90° to the piston pin—should be lightly burnished. A matted finish, as if the piston had been lapped, means that abrasive particles are present. If the cylinder bore feels like a cat's tongue, you can be sure the air filter failed. Scuffing, deep scratches on the major thrust face mean lubrication failure that may be compounded by overheating. In severe cases, the piston welds itself to the bore leaving splashes of aluminum on the cylinder. A piston with this kind of damage cannot, of course, be reused.

It's rare, but not unknown, to find a broken Briggs & Stratton piston. If the piston cracks at the skirt, it ran loose in the bore. Cracks on the undersides of the piston-pin bosses result from fatigue failure.

The ring grooves, or lands, act in concert with the piston skirt to minimize blowby by holding the rings square against the cylinder wall. In addition, the grooves limit ring upward motion that would result in flutter and early failure. Because of lack of lubrication, the upper ring groove wears four or five times more rapidly than the second groove. The oil-ring groove hardly wears at all.

As the piston climbs toward top dead center, the ring bears against the lower face of the groove. A sudden reversal of movement occurs at tdc. The ring lifts off the lower groove face, hovers for a micro-moment, and then is slammed down by the upper face of the groove. This pounding deforms the soft aluminum groove face to produce what we call wear. As the ring grows thinner in service, the chisel, as it were, becomes sharper and the groove face deforms more with each reversal of direction. This explains why wear is most severe at the outer limits of the groove.

The piston also reverses direction at bdc, but forces are smaller than those at tdc, and no appreciable groove wear results. Ring rotation—the rings spin on the piston at about 100 rpm—makes a minor contribution to the problem.

Remove the rings from the piston with a ring expander, being careful not to cut yourself. Ring edges are razor sharp. The easiest way to remove the carbon from the grooves is to farm out the job to an automotive machinist for chemical cleaning. The piston will come back looking like new. Ring-groove cleaning tools work, but tend to shave metal. Most do-it-yourself mechanics use a broken ring, mounted in a file handle, as a scraper.

Position a new ring in the upper groove and insert a feeler gauge under it as shown in Fig. 7-38. Side-valve engines can live with 0.006 in.; OHV engines are safe with 0.005 in., except for Vanguards, need a few thousandths less.

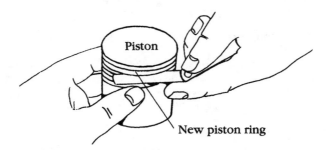

FIG. 7-38. *Determine ring-groove width with a new ring and a feeler gauge.* International Harvester Corp.

The surest way to measure piston-to-cylinder clearance is with a feeler gauge. An ordinary feeler does not reach far enough into the bore to be definitive. Wholesale Tool and other industrial supply houses supply 6 in. long leaves in the 0.001-, 0.002-, 0.003-, 0.004-, and 0.005-in. thicknesses necessary for our purposes. To gauge the wear, place the piston—without rings—into the upper bore and insert the thinnest gauge between the bore and a piston thrust face (90° to the piston pin). It should fit loosely. Repeat the operation with progressively thicker gauges to the point when a light drag is felt. This is the piston-to-bore clearance. To verify, attempt to insert a 0.001-in. thicker gauge, which should not enter or enter with difficulty. Move the piston at mid-bore, test, and test again with the piston at bdc. You can also get a fix on bore wear with a new ring, as described in the following section.

Consideration should be given to replacing the piston if the play is more than 0.004 in. More than 0.005 in. calls for a rebore and an oversized piston.

Pistons for Dura-Bore (cast-iron sleeved) engines are often tin-plated to facilitate break-in, which gives them a dull cast when new (Fig. 7-39). Many have the letter "L" stamped on the crown and several use an expander behind the oil ring. A chromed upper compression ring is standard on the better engines.

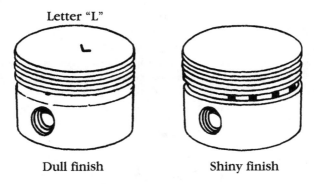

FIG. 7-39. *Cast-iron blocks and aluminum blocks with cast-iron sleeves use either tin-plated or uncoated pistons. Some tinned pistons have the letter "L" stamped on the crown, but the practice is not universal. Kool-Bore (i.e., aluminum-bore) engines require chrome-plated pistons.* Briggs & Stratton Corp.

Pistons for Kool-Bore (aluminum-bore) engines are chrome-plated and fitted with cast-iron rings. The tin-plated piston cannot be used in the aluminum bore, nor can the chrome-plated piston live in the iron bore.

Replacement pistons are available in standard and 0.010 and 0.020 in. oversizes. The oversize is stamped on the crown.

Before separating the piston from the connecting rod, note any reference marks that may be present on crown. The letters "F," "MAG," an arrow or a notch identify the flywheel side of the piston. Because pin bores are offset, incorrect assembly on the rod will result in knocking. Also make note of the conn-rod match marks, referencing them to the camshaft or some other prominent feature. Intec rods sometimes have "MAG" embossed on the flywheel side.

The recess on the end of solid piston pins should be assembled in its original position.

Using long-nosed pliers remove the two circlips, or snap rings. Intec models that secure one end of the pin with a step machined into the bore use only one circlip. Remove the piston pin. In theory, the pins float and come out with light thumb pressure; in practice pistons warp enough in service to make the tool shown in Fig. 7-40 useful. Some mechanics heat the piston with the crown down on a hot plate, while others lay the piston on wood vee blocks and drive the pin out with a small punch. If you employ this latter method, keep the punch against the pin, because if it wanders into the bearings the piston is ruined.

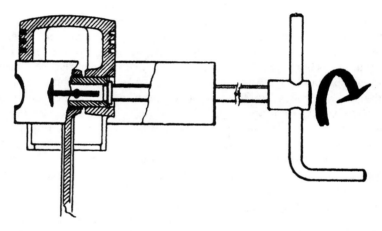

FIG. 7-40. *A piston pin press is something to look for in pawnshops and swap meets.*

The wear limit for all pin/piston bores is 0.0005 in., which is difficult to measure with the necessary precision. Any perceptible looseness in the pin fit is grounds for rejection of the pin, piston, and/or the connecting rod. Oversized (0.0005 in.) pins are available for some models, although if truth be told, engines click away for years with loose pins.

New circlips are cheap insurance and should always be used. When two clips are used, place a new clip in its groove in one piston-pin bore. Compress the clip just enough to slip into the groove. Lubricate the pin, piston, and conn-rod bearings with motor oil or assembly grease. Insert the piston pin, flat end first for solid pins, either end first for hollow pins. Press the pin through the bore, mate it with the connecting rod (observing correct rod and piston orientation) and press the rod home against the circlip previously installed or the bore step. Install the remaining retainer clip. Make certain the clips seat into the grooves. Rotate each circlip a few degrees to varify that it tracks in its groove.

Rings

Counting from the top, the first and second rings seal compression and combustion pressure. In addition to its sealing function, the second ring scrapes excessive oil from the cylinder walls, and is sometimes called the scraper ring to distinguish it from the compression ring. The oil-control ring picks up oil splashed on the lower bore and distributes it along the path of ring travel.

A piston ring is a pressure-compensating seal. When there is no pressure in the cylinder, the ring lies dormant, exerting only a few ounces of spring tension against the bore. As pressure above the ring rises, some of this pressure bleeds over the top face of the ring and, acting from behind it, cams the ring hard against the bore. The greater the pressure, the stronger the camming action and the more tightly the ring hugs the cylinder walls.

The process depends upon residual tension—the "springiness" of the ring—that holds it against the bore in the absence of gas pressure. This preload creates enough of a barrier to divert pressure behind the ring and initiate the sealing process. By the same token, the rings must be free to move in their grooves. Carbon and varnish-bound rings have ceased to function.

Figures 7-41 and 7-42 illustrate typical Briggs & Stratton ring profiles. Aluminum-bore engines use cast-iron rings; OHV iron-sleeved engines employ a chrome-plated compression ring with a convex profile. The plating develops microscopic cracks that act as oil reservoirs. The radius narrows the effective width of the face for better sealing.

Chrome replacement ring sets are available for certain of the newer engines with worn, but still serviceable, bores. Older model engines used high-tension "engineered" ring sets for this same purpose.

As can be seen from the illustration, compression and scraper rings have a definite top and bottom. Installing the rings upside down costs compression and, in the case of the scraper, increases oil consumption. The upper side is

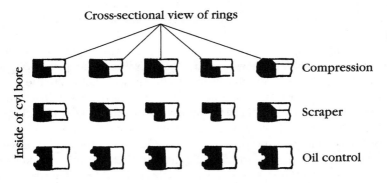

FIG. 7-41. *Cross-sections of various Briggs & Stratton rings.*

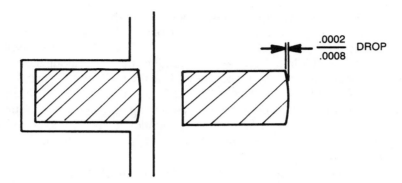

FIG. 7-42. *Radius-face, or barrel-face, compression rings are standard on the better Briggs engines. The slight radius produces high contact forces against the bore and maintains contact when the piston reverses direction at the end of each stroke.* Sealed Power Corp.

usually, but not always, identified with the letter "X," a dot, the word "Up" or "Top." Examine the old rings as you remove them to understand the geometry.

The ring gap—the distance between the ring end—is the only practical measure of ring wear (Fig.7-43). For side-valve engines with aluminum bores the wear limit is 0.035 in. on both compression rings; those with iron sleeves are set up a little tighter at 0.030 in. Because OHV engines develop more combustion pressure, the compression-ring gap should be no more than 0.020 or 0.030 in. Oil rings take care of themselves and, in any event, are replaced as part of the set.

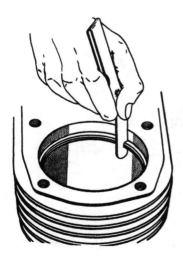

FIG. 7-43. *When a new ring is used, the ring-end gap provides an indication of bore wear.* Kohler.

The variations in ring gap at different positions in the bore serve as a poor man's cylinder gauge. Using a piston as a pilot to seat the ring squarely, establish the amount of gap at the bottom of the bore. This is the baseline, representing zero wear. Repeat the measurement at intervals along the length of the cylinder. Ring gap will increase and peak at the upper limit of ring travel. As a general rule, each 0.003 in. of gap growth represents 0.001 in. of bore wear. The difference between top and bottom readings represent taper.

The big problem with rings is hardly mentioned in the literature and rarely acknowledged by mechanics. New rings often break within the first hour or so of startup.

No one is surprised to find broken rings in a worn engine; indeed, a shattered ring justifies the mechanic's work. But replacement rings should live more than a few hours. In some cases, particularly those involving the upper compression ring, an overly wide ring groove is the culprit. But post-overhaul breakage occurs about as often when rings are fitted to a new piston. Nor can break-in stresses be blamed, since new engines seem immune to early ring failure.

The reason must be the way the mechanic handles the rings. He performs two operations: expanding the rings to slide over the piston and compressing them to enter the cylinder bore. If done incorrectly, either of these operations can compromise the rings.

Purchase or rent an expander like the one shown in Fig. 7-44. This tool pries the ring open while holding the ends on the same plane. Install the oil-control ring, together with the expander, and the two compression rings. Check that the upper rings are in their proper grooves and have their correct sides up. Stagger the ring gaps 120° from each other.

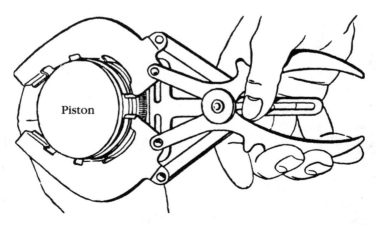

FIG. 7-44. *Use an expander to remove and install piston rings.* Briggs & Stratton Corp.

Using a brush, liberally coat the rings, piston pin and skirt, and cylinder bore with motor oil.

Small-engine ring compressors are available on the Internet and from auto-parts houses that carry the K-D line of tools. Mount the piston assembly in a vise, using wood blocks to cushion the connecting rod in the vise jaws. Tighten the compressor over the rings. Exert just enough force to overcome residual ring tension; any more force just makes installation more difficult and masks the resistance felt when a ring hangs up at the mouth of the bore.

Place the assembly in the bore so that the compressor rests squarely on the fire deck. Push the piston out of the tool and into the cylinder. Thumb pressure suffices if the compressor has been tightened just enough to overcome ring tension. In practice, one often has to resort to a dowel pin or hammer handle (Fig. 7-45). If the piston hangs, stop. A ring has escaped the compressor or the rod has jammed against the crankshaft. Withdraw the assembly and start over.

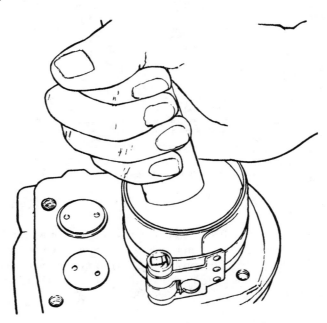

FIG. 7-45. *The piston is installed with the aid of a ring compressor, shown here being used on a Kohler engine.*

Once the piston is home, lubricate and torque the connecting-rod bolts. Turn the crankshaft through a few revolutions by hand. The piston should move without protest, slowing a bit at mid-stroke where friction per degree of rotation is greater than at the dead centers.

Cylinder bore

Some mechanics determine cylinder-bore wear by the "wobble test" and the strength of their convictions. A slightly better method is to use a ring as a gauge as described in the preceding section. But there is no substitute for a cylinder gauge or inside micrometer (Fig. 7-46). As a general rule, subject to individual interpretation, 0.003 in. of wear or 0.001 in. out-of-round mean that the cylinder should be rebored to the next oversize.

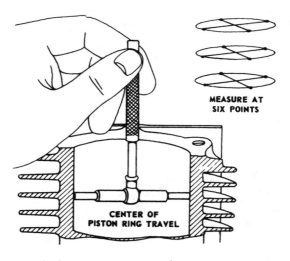

MEASURE AT SIX POINTS

CENTER OF PISTON RING TRAVEL

FIG. 7-46. *Measure the bore at six points to determine oversize, out-of-round and taper.*

Briggs supplies 0.010., 0.020., and 0.030-in. oversized piston and ring sets for most modern engines. Cast-iron blocks accept larger overbores, but finding appropriate pistons can be a problem. . Kool-Bore engines can be bored oversize and, if necessary, fitted with cast-iron sleeves. Depending upon who does it, the job costs anywhere from $75 to $150.

Boring

Most do-it-yourself mechanics would not dream of boring a cylinder oversize. But the work can be accomplished by anyone who has the patience to make the setup and access to a milling machine or heavy-duty drill press. If possible, use a Criterion or similar boring head with a graduated feed that permits the cutter to be advanced in 0.001-in. increments. You will also need to have the replacement piston in hand; at the end of the day, the piston is the only gauge that matters. Remove ball or roller bearings from the block to protect them from contamination

The instructions that follow are far from prescriptive. However you decide to go about it, the first order of business is to construct a heavy-duty holding fixture like those shown for horizontal and vertical blocks in Fig. 7- 47. The ½-in. steel plate need not extend over the entire flange: the indexing pins are the critical elements. Figure 7-48 is a working drawing for a full-coverage plate for smaller Briggs engines. Rigidity would be enhanced if the base were welded to the vertical plate and not bolted as indicated on the drawing. Overhead-valve engines with bores inclined 30° off the horizontal complicate the set up. The fire deck must be dead parallel with the machine work table, which means that the labor involved with squaring the fixture makes the boring operation anti-climatic.

Briggs & Stratton piston oversizes are based on the diameter of the original piston, and not on the bore diameter. In other words, a piston stamped "0.020" is twenty thousandths larger than the standard piston. If the bore is machined

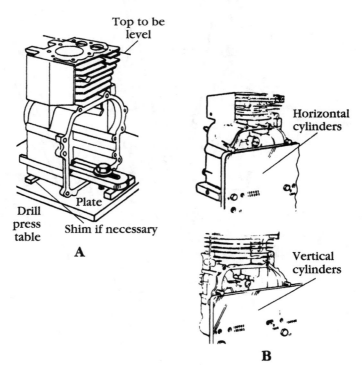

FIG. 7-47. *These drawings show holding plates for horizontal- and vertical-shaft engines. However the block is secured to the work table, the bore centerline must be parallel with the flange or bearing cover gasket surface and at 90° to the fire deck.*

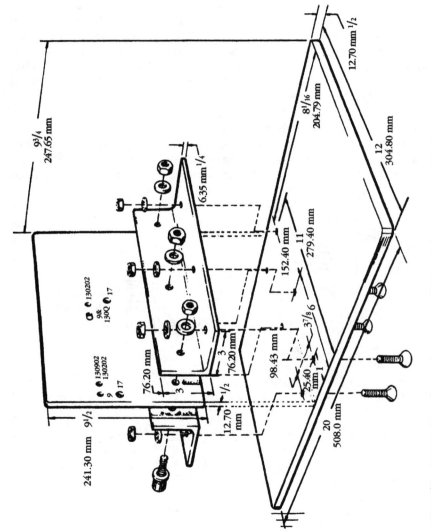

FIG. 7-48. (a) A holding fixture for most small Briggs & Stratton vertical-shaft engines.

9³/₄
247.65 mm

6.35 mm ¹/₄

8¹/₁₆
204.79 mm

12.70 mm ¹/₂

12
304.80 mm

11
279.40 mm

152.40 mm

37/8 6

98.43 mm

3
76.20 mm

¹/₂
12.70 mm

25.40
mm 1

20
508.0 mm

76.20 mm

3

130902
130202

9
17

130202

9&
130Q 17

241.30 mm   9¹/₂

168

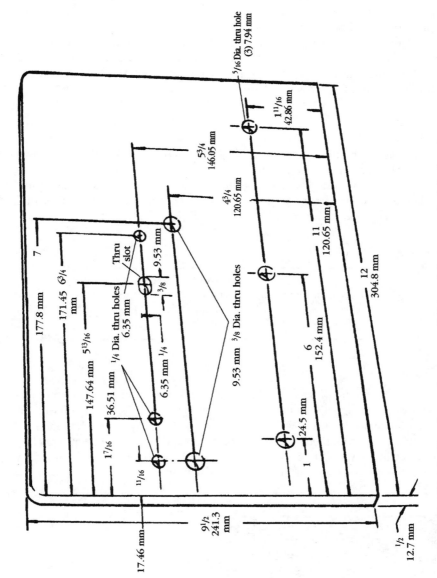

FIG. 7-48. *(b) A holding fixture for most small Briggs & Stratton vertical-shaft engines.*

169

0.020-in. over its original diameter, the new piston will have the correct running clearance. This point causes confusion, since auto and many small engine manufacturers base oversizes on the bore, which means that a 0.020-in. over piston is exactly 0.020 in. larger than the original bore diameter. For these applications, the bore must be machined 0.022 in. or so to provide clearance.

Aluminum and cast iron are friendly metals, easy to machine. Take light cuts and feed the tool at a constant rate, backing off when the cutter just peeks out from under the lower edge of the cylinder. Stop when you are within 0.003 in. of the desired diameter.

Honing

Without disturbing the setup, use an adjustable hone to bring the cylinder to size. Most shops recommend #320-grit stones for cast iron bores and iron rings, # 280-grit for chrome rings. On those rare occasions when aluminum-block engines come in for a rebore, #320 stones seem to work as well as any. Suggested speeds are:

Bore diameter (in.)	Spindle (rpm)	Strokes per minute
2	380	140
3	260	83

This combination of rotary and oscillatory speeds should produce a cross-hatch pattern at 30°–45° to the bore centerline (Fig. 7-49). Do not allow the tool to pause at the end of strokes, but reverse it rapidly. Excessive pressure loads the stones with metal fragments, dulling it and producing deep scratches on the cylinder walls. Use plenty of lubricant, such as mineral oil or an equivalent product with a viscosity of 45 SUV at 100°F.

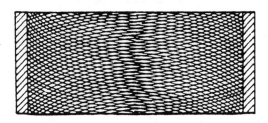

FIG. 7-49. *Preferred cross-hatch pattern.*

Allow the bore to cool for an hour or so before making the final cuts. When you are satisfied that the piston clearance is adequate for the intended service, "plateau" the bore by running the hone in a few times with #600 wet-or-dry abrasive paper wrapped around it. This procedure flattens the sharp peaks left by the stones.

A word about running clearances: engines that work at low throttle angles and that will undergo careful break-in can get by with 0.0015 in or so of clearance per inch of bore diameter. Hard usage requires more clearance to allow for thermal expansion. Racing engines are set up loose with final piston-to-bore clearances of 0.006 or even 0.007 in.

The chore that remains is to clean the bore. Never use a solvent on a honed bore: the solvent will merely float abrasive particles deeper into the metal where they remain. Instead, scrub the bore with a brush, hot water, and Tide or another strong detergent. Rinse and dry with paper towels. Repeat the process until the suds are white and the paper towels show no discoloration. Oil immediately.

Glaze breaking

Traditional practice is to roughen the glass-smooth glaze of compacted iron crystals formed by the rubbing action of the rings on used cylinders. Some mechanics, with the blessing of Perfect Circle, omit this step. While everyone agrees a roughened bore has no effect upon the rate at which chrome rings seat, most small-engine manufacturers continue to recommend glaze breaking for cast-iron rings. It hardly seems worthwhile to hone aluminum blocks, but nothing will be lost by doing so.

The process is the same as for honing after boring, except that the aim is to merely roughen the surface. Use a #320-grit hone as described earlier and, when finished, scrub down the cylinder with detergent and hot water.

Pushing the envelop

Briggs & Stratton recently dedicated 10,000 sq. ft of factory floor space to assembling $2000 racing engines that, for many kart racers, are little more rough castings. Some of the techniques developed for competition can make utility engines more durable and, in the process, works of high craftsmanship.

Modifications center on the bore, which should remain round and true under thermal and mechanical stresses. It's not unusual to pull a motor down and see patches of the original cross-hatching in the otherwise smooth cylinder bore. Nor it is uncommon to sense a certain dulling of performance as an engine reaches operating temperature and the rings no longer seal.

Thermal stresses don't receive much attention from the racing fraternity, probably because they are difficult to simulate during machining. But mechanical stresses generated by head and flange bolts can be replicated easily enough.

In so far as possible, all machining operations are carried out with the flange casting and a torque plate assembled to the block. The torque plate consists of a ½-in. thick steel plate, surfaced on both sides, and with the same bolt pattern as the head. A hole in the center of the plate gives access to the bore. Torqueing the plate down distorts the bore in the same manner that the cylinder head does. In addition, the almost imperceptible bulges in the cylinder walls created by the head bolts are duplicated. Some mechanics wait a day or two before proceeding to allow the stresses generated by the plate and flange to relieve themselves. When the block is bored and honed, it will retain its shape.

Main bearings

Briggs & Stratton engines use plain or anti friction (ball) bearings. When the latter is specified, either of two arrangements is used to secure the bearing. Modern engines employ press fits on both the inner and outer bearing races; earlier models use a press fit at the inner race/crankshaft and pin the outer race to a carrier that, in turn, bolts to the block.

Two types of plain bearings are used: DU Teflon-impregnated bronze bushings for the better engines, and block metal on the less expensive models.

The condition of ball bearings cannot be gauged, and must be judged by look and feel. Soak the bearing in solvent to remove all traces of lubricant and allow it to air dry. If you use compressed air to speed the drying process, the air line must be fitted with a water trap, since these bearings have zero tolerance for moisture. Turn the bearing by hand. It should roll without catching and without the cracks and pops associated with brinelled faces and pitted balls. Some outer race play is allowable, although it should be limited to a few thousandths of an inch.

Ball bearings come out with crankshaft or remain in the block and/or flange. The bearing carrier arrangement used on older side-valve engines frees the outer races, but leaves the bearings on the crankshaft. Normally these bearings are not disturbed. If replacement is necessary, remove the old bearings with a bearing splitter (Fig. 7-50) and install the new bearings with a length of pipe sized to match the *inner* race (Fig. 7-51). Pto-side bearings on OHV engines slip over the crankshaft and remain in the flange. Remove and install these bearings with a tool sized to fit the *outer* race. If you have access to a lathe, fabrication of the driver should present no problems. Otherwise, the cost of the factory tool makes this a job for an automotive machinist.

DU bushings also require a special driver that in the factory version extends into the counterbore beyond the bushing to function as a pilot. When present, the oil hole in the casting should index with the hole in the bushing. Some bushings are seamless, others are rolled from sheet stock and

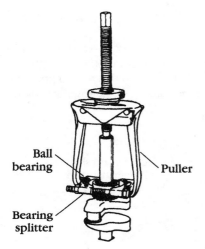

Ball bearing

Puller

Bearing splitter

FIG. 7-50. *Ball bearings are extracted from the crankshaft with a bearing splitter.* Tecumseh Products Co.

FIG. 7-51. *When bearings are driven home, use a tool sized to the inner race and support the crankshaft webs with a wood block.* Rockwell Mfg. Co.

have a split where the ends come together. The split should be well clear of the slots provided for peening. Install to the depth of the original. With a little ingenuity, DU bushings can be adapted to plain-metal bearing crankcases and to flanges that have enough metal surrounding the boss to accommodate the bushing OD.

Most plain-bearing blocks can be reamed to accept a brass or DU bushing, as can many flanges. The special and expensive tooling required to ream bearings gives Briggs & Stratton dealers a virtual monopoly on the work.

Magneto-side seal

The crankcase, or magneto-side, seal is removed and replaced as described for the pto seal in the Flange section.

Crankshaft

The crankshaft is the fundamental part upon which all else depends. Because it has nothing directly to do with performance, this gangly piece of iron is often taken for granted. But flaws in the crankshaft mean that the engine will not live, and all the polishing and fitting that go into a rebuilt engine amount to naught (Fig. 7-52).

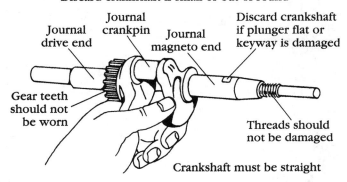

Discard crankshaft if small or out of round

Journal crankpin

Journal drive end   Journal magneto end   Discard crankshaft if plunger flat or keyway is damaged

Gear teeth should not be worn

Threads should not be damaged

Crankshaft must be straight

FIG. 7-52. *Crankshaft inspection points. Magnetron shafts do not have the flat milled magneto journal.* Briggs & Stratton Corp.

Begin inspection with the magneto-end threads that, if crossed or stripped, cannot be trusted to anchor the flywheel against engine torque. Conventional wisdom holds that the keyway must be true and square in order to maintain correct ignition timing. The factory-recommended solution for a wallowed keyway is a new crankshaft.

Yet, when you think about it, the key is no more than a marker for assembly. The strength of the flywheel/crankshaft connection, i.e., the integrity of the timing adjustment, depends upon the taper and the torque applied to the starter nut. It has nothing to do with the shear strength of the aluminum key. Kart racers sometimes discard the key in order to make timing changes easier.

A crankshaft with a wallowed keyway can be reused if a new key is centered properly within the keyway. Normally only one side of the keyway will be deformed and the key can be installed against the undamaged side. Mount the flywheel, tap it down over the crankshaft taper with a mallet to fix the key in position, and tighten the flywheel nut to specification. For added security, you may want to roughen both tapers with sandpaper, swab down the mating surfaces with alcohol, and apply three or four drops of red Loctite before tightening the nut. Ignition timing may be off a few degrees, but the engine should run and start without problems.

Modern Briggs crankshafts are made of a special alloy that, according to a company engineer, has made heat-treating redundant. However, the cranks for some John Deere engines are rolled after grinding to close the grain structure for better surface finish and durability. In the past, heavy-duty crankpins were heat-treated. These cranks can be recognized by the dull black edges around the pin journal.

Small imperfections on the bearing journals polish out, but deep, nail-hanging grooves mean that the crank must be replaced together with the associated bearing. As a general rule subject to the intervention of real-world concerns, such as cost and parts availability, the wear limit is 0.0010–0.0013 in. for any of the three journals on plain-bearing engines. Journals that run on ball-bearings are immune to wear, unless the bearing has spun. Most mechanics judge crankshaft condition by feel. The journals should fit plain bearings with just perceptible play, representing something between 0.001 and 0.002 in. Excessive play is almost always the fault of the plain bearing and not the crankshaft.

The critical surfaces are at the interface between the crankpin and connecting rod. This bearing sees three times as much wear as the others, and failure can send the broken rod crashing into the block. If there is no obvious damage—bluish heat discoloration, deep scores or flecks of aluminum welded to the pin—check the clearance with plastic gauge wire as described under "Connecting rods."

The factory supplies 0.020-in. undersized rods for many engines, together with instructions for the crankshaft grinder. If you go this route, shop around for a reputable machinist and check his work against the instructions. The fillets should be rounded to reduce stress concentration and bearing clearance exactly as specified.

Briggs, like other small-engine manufacturers, holds crankshaft straightness to within ±0.001 in. and adamantly opposes attempts to restore this dimension. While the company is, no doubt, happy to sell crankshafts, the impetus for this policy is the legal vulnerability that straightened crankshafts present. If the shaft cracks, and the cracks may not be visible, the shaft will part in service. It doesn't take much imagination to picture what could happen if the shaft is attached to a rotary lawnmower blade. But automotive machinists routinely straighten shafts that are a few thousandths out of true. The choice is up to you and your machinist.

8

The Europa

Many readers are familiar with flathead motors, but might be a little shy of the newer valve-in-head models that represent a fundamental change in the way Briggs makes engines. Consequently, a 5.5-hp vertical-shaft Europa 97700 was purchased, taken apart, and photographed. What remained was to determine if the Europa would hold up in service. The spindly pushrods and soft valve springs, not much more resilient than rubber bands, looked like a prescription for swallowed valves. The report at the end of this chapter describes how it was used to clear jungle growth in Mexico.

This particular engine turned out to be a good subject for investigation because its valve gear became standard on Intec and other Briggs OHV models. Like the Model P that fixed the architecture of utility engines through the 1940s, the Europa was a seminal design.

The name "Europa" came about because the shape of the plastic shroud was said to reflect European styling. But that was the limit of foreign influence: the Europa was a homegrown product, built at the main plant in Wauwatosa, Michigan. While the Walbro carburetor, Magnetron ignition, revolving oil slinger, and block-metal bearings are carryovers from side-valve engines, other features, such as the concave piston that forms the bulk of the combustion chamber, the four-bolt head, and the Eaton-type oil pump were novel. "We had," said one of the engineers involved in the project, "a blank sheet of paper."

The engine found a limited market as power for generator sets and upscale walk-behind lawnmowers, but it's relatively high price limited sales. Listing at $317, the Europa cost some $70 more than the 4-hp I/C 114900 and twice as much as the Standard version of that same 11-cid engine. OEM customers, of course, paid less, but still found the Europa an expensive proposition.

Technical description

Initially, Briggs announced plans for a whole series of Europa 147-cc engines, but only two models—97700 and 99700—were built. The plastic camshaft, resilient valve springs, in-piston combustion chamber, and semipressurized oiling system would be incorporated in subsequent OHV engines.

Valves mount vertically above the piston and actuate through pushrods and stamped-steel rocker arms. The crankshaft rides on block metal, a practice that Briggs engineers defend on the grounds that 90% of customers discard an engine before the main bearings fail. If necessary, the block can be reamed and fitted with bushings.

Two oiling systems are used, which calls to mind the old 216.5-cid Chevrolet that had scoopers on the connecting rods, squirters in the pan, and a pump to lubricate what was missed. Splash provides the Europa with primary lubrication, by means of a paddle wheel driving off the camshaft gear. An eccentric oil pump, also camshaft-driven, delivers pressurized oil to the upper main bearings and rocker arms. A technician involved in Europa development reported that the provision for upper main-bearing lubrication was inspired by Tecumseh practice.

As always, the connecting rod runs directly against the crankshaft, without the benefit of replaceable insert bearings. The crankpin can be reground to accommodate 0.020-in. undersized rods.

A composite plastic and steel camshaft is used. Plastic cam lobes have become a Briggs trademark. Although the cams work, the trade-off involves light valve springs and severe limits on valve acceleration and lift. Europa valve springs can be compressed with one finger.

Selectively fitted pistons, graded by size in increments of a hundredth of a millimeter (0.00039 in.), run in iron liners that appear to have been centrifugally cast for grain uniformity. Combustion chambers receive nearly 100% machining to eliminate casting inaccuracies. Until recently, machined chambers were confined to diesel engines. Sintered iron bearings, which combine extreme hardness with porosity for oil retention, locate the rocker arms.

The Europa required a dedicated transfer line and was expensive to produce. The number of parts was held to a minimum. Except for the valves themselves, all intake and exhaust components interchange. Bolt holes in the flange are numbered for the correct assembly sequence. A slight reduction in the pto end of the crankshaft allows the flange to be assembled without risk of damage to the lower oil seal.

Major specifications are listed in Table 3-1.

While horsepower data is incomplete and sometimes contradictory, the Europa appears to hold the record as the most potent engine for its size that Briggs ever built. Developing 5.5 hp from 8.97 cid, the engine is clearly superior to side-valve models that displace as much as 3 in.^{3}/hp.

Table 3-1
Europa general specifications

Bore	65.06 mm
Stroke	44.20 mm
Displacement	147 cc
Connecting-rod length	76.2 mm
Compression ratio	8.5:1
Initial spark timing	22° before tdc
Intake valve opens	320° after tdc
Intake valve closes	104° before tdc
Exhaust valve opens	96° after tdc
Exhaust valve closes	328° before tdc

A closer look

Table 3-2 lists specifications that mechanics should be aware of. The only departure from conventional B & S specs is the valve lash, which has been tightened to allow for thermal expansion.

Daniel Faulkner, a Houston-based photographer specializing in auto racing and industrial work, shot the photographs that illustrate this section. For readers interested in the techniques of photographing machine parts, Mr. Faulkner used an F-4 Nikon with a Nikon 55-mm Micro-Nikkor lens, an Apollo soft box, and Norman flash equipment. Seamless white paper served as a backdrop.

Disassembly

Follow this procedure:

1. Some models come with a fuel cutoff valve. If your engine does not have this feature, drain the tank by disconnecting the hose at the carburetor fitting.
   Warning: Perform this operation outdoors in an area remote from potential ignition sources. As an additional precaution, place the control lever in the stop position.
2. Remove fuel tank assembly (integral with the decorative plastic shroud), oil-filler tube, and muffler guard. Note the plastic spacer between the carburetor-side, shroud-mounting lug and engine block.
3. Remove the three-piece blower housing and attached starter.
4. The combination of 60-lb/ft-assembly torque on the flywheel nut and an aggressive taper fit means that the flywheel must be pulled, rather than knocked loose. Unless you have access to a factory puller, which uses self-tapping screws, it will be necessary to

Table 8-2
Europa 97700 99700 service specifications (inch)

Standard clearances:	
Spark-plug gap	0.030
Armature air gap	0.007–0.013
Valve clearance	
Intake	0.005
Exhaust	0.006
Reject dimensions:	
Crankshaft journals	
Magneto side	0.873
Crankpin	1.122
Pto side	1.060
Cam bearing journals	
Magneto side	0.615
Pto side	0.0615
Bearing bosses	
Magneto-side main bearing	0.878
Flange-side main bearing	0.1.065
Magneto-side cam bearing	0.621
Flange-side cam bearing	0.621
Torque limits:	
Flywheel nut	60 lb/ft
Cylinder-head bolts	165 lb/in.
Flange bolts	90 lb/in.
Connecting-rod bolts	105 lb/in.
Electric starter hold-down bolts	85 lb/in.
Governor lever lock nut	35–45 lb/in.

tap $1/_4 \times 20$ TP1 threads on the two holes cast into the hub, as shown in Fig. 8-1.

5. A suitable puller can be fabricated from bar stock with bolt holes on 1.5-in. centers (Fig. 8-2).

6. The Europa purchased for this exercise was a "compliance engine," meaning that it was fitted with a spring-loaded flywheel brake and ignition-shorting switch. The brake assembly can be unbolted or "cocked and locked" (Fig. 8-3).

7. Remove the air cleaner assembly, which is secured by three $5/_{16}$-in. capscrews.

8. Use long-nosed pliers to unhook the throttle spring at the carburetor (Fig. 8-4).

9. Disengage the small spring that actuates the choke (Fig. 8-5). Although the spring is delicate and might rust, it is a major improvement over previous choke-engagement mechanisms.

FIG. 8-1. *Flywheel hub holes receive threads in preparation for disassembly.*

FIG. 8-2. *A homemade puller works about as well as the factory tool and costs nothing to fabricate.*

10. Once the springs and ignition-kill wire are disconnected, gently rotate the carburetor assembly to release the governor link (Fig. 8-6).
11. Remove the valve cover, rocker-arm nuts, rocker arms, plastic valve spacers, and pushrods (Fig. 8-7).
   Caution: Set screws prevent rocker-arm balls from vibrating loose. These screws must be loosened before attempting to back off the balls.

FIG. 8-3. *A screwdriver blade locks the spring-loaded brake out of contact with the flywheel rim.*

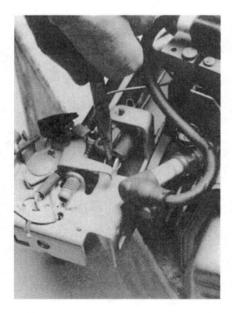

FIG. 8-4. *Long-nosed pliers are used to disengage one end of the throttle spring.*

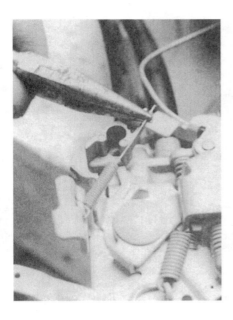

FIG. 8-5. *The choke spring looks like the sort of part that will need frequent replacement. Order PN 262749.*

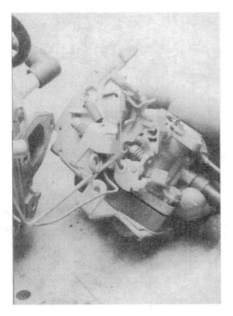

FIG. 8-6. *The carburetor body rotates out of and into engagement with the governor link.*

FIG. 8-7. *Partially disassembled cylinder head, showing the quality of foundry work now obtainable for a utility engine.*

12. Remove the four cylinder-head bolts (Fig. 8-8). While not mandatory, loosening these bolts in stages, working across the diagonals, reduces the risk of warping the head.
13. The oil pump is accessible from outside the flange (Fig. 8-9). Use a small screwdriver to pick out the impeller and rotor.

FIG. 8-8. *Four long bolts secure the cylinder head, which is stiffened by its compact dimensions and large cooling fin area. Comparable side-valve heads require eight bolts and are by no means immune to gasket failure. The arrow stamped on the piston crown is an assembly reference.*

FIG. 8-9. *The oil pump drives off of the camshaft to lubricate the magneto-side main bearing and cylinder head components. Pressurized lubrication should prolong the life of the upper main bearing, although this has not been proven. Tests of competitive engines with and without this feature show no measurable difference in bearing wear.*

14. All new engines are contaminated to some degree with chips and drill dust. The example Europa was no exception. Most of the swarf was lodged at the oil-pump inlet, although some was distributed inside the crankcase. The factory apparently "motors" engines prior to shipment, rotating the crankshaft with an external power source.
   Note: Purchasers of these fairly expensive products would do well to clean the pump and crankcase before initial startup.
15. Using strip abrasive, thoroughly clean the crankshaft outside diameter (OD) in preparation for flange removal.
   Caution: Failure to thoroughly clean the crankshaft damages the oil seal and lower main bearing.
16. Undo the flange holddown cap screws and separate the flange from the block casting with a soft mallet. Carefully withdraw the flange, which should slip easily off the crankshaft. The flyweight/oil-slinger assembly might drop into the flange (Fig. 8-10) but the camshaft should remain suspended in its magneto-side bearing to establish the original positions of the timing marks.
17. Check the timing as an assembly reference (Fig. 8-11). Unlike other engine makers, Briggs & Stratton does not index timing mark alignment with top dead center (tdc) on the compression stroke.
18. Withdraw the camshaft and tappets (Fig. 8-12). The camshaft assembly includes a compression release that holds the intake valve slightly open during cranking.

FIG. 8-10. *A camshaft-driven slinger provides primary lubrication in traditional Briggs fashion.*

FIG. 8-11. *Mechanics soon learn to verify original valve timing before the camshaft is disturbed.*

FIG. 8-12. *The Europa employs a composite plastic and steel camshaft that, like the fabled bumblebee, should not fly. In auto engines, camshaft-lobe pressures approach 50,000 psi at idle. Plastic lobes obviously work, however, as witnessed by the millions of Quantum engines that use them.*

19. As always, become familiar with the rod-cap/shank orientation marks before proceeding with disassembly. The arrow on the cap points to the camshaft and aligns with a second arrow on the upper rod assembly (Fig. 8-13).

FIG. 8-13. *Connecting-rod match marks take the form of arrows pointing toward the camshaft. When oriented properly, the rod cap snaps onto the rod shank rabbet.*

20. Withdraw the piston and upper rod assembly. Replaceable spring locks (circlips) locate the piston pin, which at room temperature should be a light interference fit with the piston. If further disassembly is required, warm the piston to approximately 200 degrees Fahrenheit with a heat gun or place it (suitably protected) on an electric hot plate for a few seconds. The piston will expand enough to release the pin. **Caution:** Driving the pin out cold can distort the piston.

Inspection

Clean parts with Varsol or an equivalent solvent. If a pressurized solvent gun is not available, submerge the block casting (less Magnetron and carburetor) in solvent for a few hours and blow out the oil passages with compressed air. Repeat until the discharge is clean.

Examine all friction surfaces for wear (Table 8-1) and surface finish. Lubricate with liberal amounts of motor oil (Fig. 8-14) or with engine-assembly lubricant. Fasteners should, if Briggs follows the earlier practice, be lightly oiled.

Obtain the replacement parts you need, which will include whatever gaskets have been disturbed, a spark plug, and air filter cartridge (PN 494586). Note that crankshaft-related parts vary somewhat between type numbers and that there was a running parts change involving the cylinder-head casting and related components. Existing documentation is not entirely clear on the point but it appears that the change occurred on August 8, 1992, or on

FIG. 8-14. *In the absence of factory recommendations to the contrary, prudent mechanics should renew the connecting rod bolts upon disassembly. Torque to 100 lb/in.*

build date 92080500. Because of differences in the head gasket, two over-haul gasket sets are cataloged: PN 496055 after 92080500 and PN 494963 on and before 92080400.

Assembly

Follow this procedure:

1. According to Briggs engineering, worn main bearings can be reamed to accept bushings. Use the tooling developed for earlier engines and readily available to dealer mechanics.
2. Replace oil seals (magneto PN 299819, pto PN 399781) as described in chapter 7.
3. Ideally, the crankshaft should be sent out to an automotive machinist for a thorough inspection and polishing At the minimum, mike bearing surfaces and look for bends outboard of the pto bearing.
4. Replacement piston assemblies include ring sets, spring locks, and piston pins, which can also be purchased as individual items:

Size	Piston Assy.	Ring Set	Spring Lock	Pin
Standard	493781	493782	262514	262514
0.010 O/S	495267	495268	262514	262514
0.020 O/S	495269	495270	262514	262514
0.030 O/S	496271	496272	262514	262514

If the piston pin has been disturbed, heat the piston as described in #20 under the previous section "Disassembly." Using new spring locks, assemble the piston, pin, and rod. Verify piston-to-rod orientation. If incorrect, the engine will knock.

Note: Connecting-rod assemblies are available in standard (PN 493049) and 0.020-in. undersizes (PN 493689) for use with reground crankshafts.

Caution: Without factory documentation, the wisest course of action is to renew the rod bolts (PN 94404) upon assembly.

5. Using a pump-type dispenser, flood the rod bearings and crankshaft journal with oil. Verify that both connecting-rod arrows point toward the camshaft and assemble the cap to the upper rod. Listen for a click when the cap seats.
6. Lightly lubricate new rod bolts with motor oil and torque down evenly in three increments to 100 lb/in. (Fig. 8-14). Turn the crankshaft through two full revolutions to detect possible binding.
7. Insert valve tappets into their respective bores. (Tappets interchange but should be returned to their original places on used engines.) Some mechanics apply assembly grease to the tappets and cam lobes

as insurance against scuffing. Grease also prevents the tappets from falling out as the camshaft is installed.

8. Inspect the crankshaft key (PN 94388) for damage. Install the timing gear over the end of the crankshaft with the timing mark visible.

9. Install the camshaft, turning the crankshaft as necessary to index timing marks.

10. The slinger/governor hangs off the camshaft stub with the paddle-shaped governor lever resting on it and positioned over the recess cast into the floor of the flange (Fig. 8-15).

FIG. 8-15. *Critical wear sites include the oil pump, camshaft lobes compression release, and camshaft gear teeth. Tappets should be returned to their original bores and liberally coated with assembly lube.*

11. Position a new flange gasket on the crankcase casting. Because the factory supplies only one gasket (PN 2721324), crankshaft endplay can be ignored. A thinner gasket could be used on earlier engines to compensate for thrust-bearing wear.

12. Mount the flange over the crankshaft stub, pressing it home with the palm of your hand.

13. Progressively torque flange screws in the sequence indicated on the casting to the 85 lb/in. torque limit (Fig. 8-16).

14. Oil pump wear limits have not been published, but new engines appear to be set up with 0.002 in. between the rotor OD and recess (Fig. 8-17). Because of its location at the bottom of the sump, the pump cover appears vulnerable to abrasion. The pump rotor and drive gear are listed under PN 493884. The cover and O-ring under PN 4927656.

FIG. 8-16. *Flange screws should be lightly oiled and torqued in the indicated sequence to 85 lb/in.*

FIG. 8-17. *Most pump wear should occur on the cover, which can be replaced inexpensively.*

15. The cylinder-head gasket (currently PN 272314; PN 272488 before 92080500) mounts either side up. Head bolts should be lightly oiled and torqued in an X-pattern to 160 lb/in.
16. Experienced mechanics judge valve guide condition by eye: a bell-mouthed guide allows the valve to wobble in the fully open position. More precise methods require special factory gauges.

Without these tools, entrust valve guide work to a competent automotive machinist, who should also be able to approximate factory valve-face angles.

17. Install the valves, noting that the intake valve assembly includes an umbrella-type oil seal (PN 493661) and seal gasket (PN 272376). No special tools are needed because the interchangeable valve springs are soft enough to compress by hand.

18. Install the rocker assemblies in this sequence: pushrod guide plate (marked *top*), pushrods, plastic valve caps (PN 262499), rocker arms, and ball nuts. Apply assembly lube to the rocker-arm/ball-nut contact surfaces.

19. Rotate the crankshaft as necessary to bring the tappets on the heel of the camshaft lobes and, on the intake side, out of engagement with the compression release. Set valve lash to the 0.003- to 0.005-in. cold specification (Fig. 8-18). Tighten the ball-nut set screws to fix the adjustment.

20. Install the flywheel key (PN 222698), flywheel, starter hub, and flywheel nut. Torque the nut to 60 lb/ft.

FIG. 8-18. *Static valve adjustments are made on a cold engine with rocker-ball set screws backed off and tappets on the cam-base circle. Make certain the intake tappet rests in its fully retracted position, clear of the compression release. Tightening the rocker-ball set screws against their respective studs fixes the adjustment.*

21. Complete the assembly by installing the carburetor, muffler, and miscellaneous parts.
22. The air cleaner assembly requires special mention. It consists of a foam precleaner and a paper cartridge, sealed by pliable gaskets and contained in a plastic housing. A careful inspection prior to disassembly revealed no obvious leak paths between the air cleaner and the carburetor intake.

Upon installation, an $1/8$-in. gap remained between the bottom of the carburetor air horn and the air cleaner mounting surface—even with the hold-down bolts torqued to the shear point. The carburetor flange threads had been cut with a tap worn undersized. Chasing the threads with a 10-32 tap corrected the problem (Fig. 8-19).

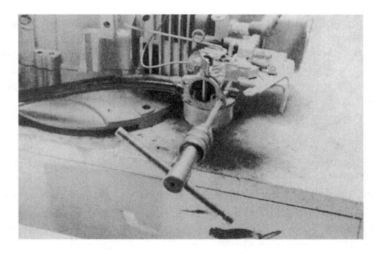

FIG. 8-19. *It was necessary to retap the air cleaner-to-carburetor-screw holes for the example engine.*

How many Europas came off the line with impossible-to-retighten air filter screws is anyone's guess. Reported tales of catastrophic failure associated with dirty air cleaners might have a basis in fact. It is also true that a leaking Europa will run to self-destruction with a clogged air filter.

Service history

The Europa was mounted on a 21-in. rotary mower, and run for more than 500 hours, with half of that time clearing construction sites in eastern Mexico. Typically, these sites are covered with heavy vegetation that grows more than 3 in. a day during the rainy season. We hack out the worst of it with machetes, then follow up with the mower. *Escumbro,* the detritus left by the construction workers, litters the ground. Collisions with chunks of concrete, rebar and half-buried roots are inevitable. About all one can do is throttle down to minimize the damage.

Once the *selva* is cleared, we plant native grasses and avocado, banana, mango, and other fruit trees provided as part of the Mexican government's reforestry program.

The environment is hard on machinery: within sight of the sea, five months of daily rain followed by dust storms, and workers who live as their ancestors did in centuries past. A Kohler 8M failed in less than 40 hours; another Kohler had its rewind starter pulled out by the roots. Without a lathe and milling machine, it would be impossible to keep the equipment running, since there is very little by way of parts support.

The ancient Southland mower was fitted with high-flotation pneumatic tires to better cope with the terrain and the mud. A local welder, an artist with an acetylene torch, periodically repairs the cracks in the cast-aluminum deck. Three blades and one blade adapter have been replaced so far.

As might be expected, vibration and wheel rumble are present at wide throttle angles. The crankshaft has a slight bend, but without parts, all that one can do is to have a machinist straighten and Magnaflux it. Since I don't trust the construction hands to operate the machine, the risk of using a straightened crankshaft is acceptable.

When all is well, the engine starts on one pull, hot or cold. Once running, it produces no visible smoke and uses just enough oil to be assured that upper cylinder lubrication is adequate. Crankshaft float has increased somewhat, probably due to blade thrust.

The only real problems have been with the carburetor. The Walbro, because of the location of its main jet at the very bottom of the bowl and its intricate low-speed circuit, clogs easily. Running the low-speed needle out to the end of its travel sometimes enables the engine to keep running, but the rich mixture soon fouls the spark plug. We purchased a second Walbro in the States and swap them back and forth, always keeping a clean carburetor in reserve. Installation of an inline fuel filter helped, but the carburetor jets remain vulnerable. How much of this is due to Pemex gasoline is difficult to say.

Viton-tipped inlet needles have been replaced many times, but tend to dribble fuel when the engine is parked. All gravity-powered fuel systems should be equipped with a shut-off valve at the tank, if only to eliminate the potential for fire.

The combustion chamber collects heavy accumulations of carbon, which take the edge off of performance unless cleaned once or twice a year. These deposits seem normal for small engines in this country and, almost surely, have to do with the fuel. A few years ago, the engine experienced a bout of bent pushrods caused by valve sticking. Now we buy gasoline only in the quantities needed for the job and these problems have not recurred.

Even when the carburetor is in tune, the Europa eats spark plugs, spark plugs that function perfectly well in an automobile or in an antique Ford tractor with battery-and-coil ignition. The Magnetron does not put out enough cranking voltage to fire a damp or normally dirty plug. Narrowing the armature air gap and substituting a known-good Magnetron did not help.

Because the straight 20W-oil Briggs recommends is not available in Mexico, we used 30W-Castrol, which is changed every 15 or 20 hours.

Unfortunately, I failed to monitor oil temperature, but researchers at the University of Wisconsin-Madison provided some data in the course of making exhaust emissions tests on a Europa 97700.[1] The engine was mounted on a dynamometer and run for 30 minutes at 3060 rpm under constant load. At half load, the oil temperature reached 100°C (212°F); at full load, i.e., the load that absorbed all available horsepower, oil temperature increased to 121°C (248°F) Based on these figures, oil temperature does not seem to be a matter of concern.

In summary, the Europa has survived a decade of hard use, far more severe than would be encountered in a suburban setting. The valve gear functioned perfectly, except when sabotaged by stale gasoline. As far as major mechanical components are concerned, it should be good for another ten years. But, like other small engines, the Europa requires frequent oil changes, and more-or-less continual attention. Normally it starts on one pull; if three pulls aren't enough, reach for the wrenches.

---

[1]F.A. Jehlik and J.B. Ghandi, "Investigation of Intake Port Films in a Small Air-Cooled Engine," SAE 2001 01 1788/4211.

9

Winds of change

Briggs fans were not happy to hear that the company would consign production of its top-of-the-line Vanguard engines to the Japanese. At the time—it was back in the mid-1980s—the Japanese seemed invincible. It appeared to many that the Japanese would destroy American small engine manufacturers in the same way they had destroyed the British motorcycle industry a decade earlier.

Briggs & Stratton's Japanese partners had good credentials. Mitsubishi, which was assigned responsibility for producing single-cylinder Vanguards, was a long-time aircraft and marine engine manufacturer. Daihatsu was given manufacturing rights to vee-twin Vanguards. In addition, Daihatsu would market a three-cylinder diesel of its own design under the Briggs name. The firm was one of the first in the Orient to acquire a license to build diesel engines and, in recent years, has benefited from a connection with Toyota.

The Vanguards proved themselves to be reliable, but pricey, engines, and no threat to Briggs core market. Probably not a single American job was lost as a result of the alliance.

The threat poised by Honda was less easy to contain. The company had a history of sacrificing earnings to gain market share, which was a strategy that Briggs, tied closely to its quarterly earnings reports, could not easily counter. Honda's small engines enjoyed a "halo effect" from the company's superbly engineered automobiles and motorcycles. By designing and manufacturing its own lawnmowers, Honda exerted direct control over quality and the "feel" of the product. Most of Briggs production was sold to OEMs. Honda also launched a nationwide ad campaign showing American workers assembling Honda mowers.

Briggs' response was to concentrate on the product, which is the same advice Nelson Mandella gave Clinton when he was in trouble. The most

intensive engineering effort in the company's history resulted in two gen-
erations of overhead-valve engines. Many of the old-line "heritage" motors
were dropped and those that remained were updated. Briggs also stream-
lined manufacturing, outsourcing pistons, rewind starters, and components that were formerly made in-house. Millions were invested to update
U.S. manufacturing plants, including $5 million for a battery of CNC crank-
shaft lathes at the Poplar Bluff, Missouri, facility. This new equipment
turns out a crankshaft every 18 seconds, machining both ends with the
same setup.

The effort paid off. As one company executive put it, "Our engines are at
least as good as Honda's—which are overrated—and generally better." The
executive in question is a former motorcycle racer who knows the hardware. He cited the results of exhaustive, one-on-one tests run at the Florida
facility to prove his point. Briggs is so confident of its ability to compete with
arch-rival that it fits Honda engines to its gen-sets, if the OEM customer asks
for them.

Briggs continues to hold its lead in the U.S. market, with domestic sales
of some 7.5 million units a year. Honda sells a third as many (2,227,000 units
in 2006). As far as the writer knows, this is the only time that an American
manufacturer of mid-tech products beat the Japanese at their own game.

The company faces another challenge, which has more to do with manufacturing costs than technology. The Chinese are coming. In fact, one model
of the MTD mower comes with a Chinese engine. According to a company
spokesman, all that's holding the Chinese back are the high warranty costs
associated with their lack of quality control. But QC can be learned.

The company responded by shifting some of its production to China.
Briggs & Stratton Changquing manufacturers cast-iron block, single-cylinder
10- and 16-hp engines in the city of that name in southwestern China.
These are the same workhorses that Briggs stopped making in this country
because of emissions requirements. Plans are to add aluminum blocks to
the production mix that should reach between 800,000 and one million
units a year.

By acquiring a Chinese partner, Briggs has almost guaranteed itself a
place in what could become the world's largest market. The Changquing
factory could also export low-cost engines to the United States, but the logis-
tics are complicated and there are no immediate plans to do so. Briggs has
also opened a plant in the Philippines.

Construction is underway on a factory in Czechoslovakia. Again, the
appeal seems to be low labor, rent, and pollution-control costs coupled with
ground-floor access to a developing market. Meanwhile, the company has
made massive new investments in its Milwaukee plant, where operations
began almost a century ago.

Clean air

There's something restful about the simple, repetitive work of cutting grass. The smell of fresh-cut grass, the hum of the mower, and the warmth of sun takes us back to a place where we belong. Old people worry about the loss of driving privileges; for me, the incapacity that comes with age will mean that it is no longer possible to walk behind the mower.

The small engines that have made this and thousands of other useful tasks possible also impose environmental costs. That no one can deny. The argument is over the degree of culpability these engines have and the severity of the regulations necessary to clean them up.

As world's largest manufacturer, Briggs finds itself at the center of the controversy. Both the regulators and the company have muddied the issue with half-truths and exaggerated claims. To understand what's going on, it is necessary to first talk about who the regulators are and what it is that they object to.

The clean-air forces

Two regulatory bodies oversee air quality in the United States—the federal Environmental Protection Agency and the California Air Resources Board (CARB). Historically these agencies have worked together, with CARB setting exhaust-emission rules for California and the EPA following up with similar, but less stringent, rules for the rest of the country. Other states could choose between CARB or EPA rules, but choose they must, or forfeit federal highway funds. By the end of the century, both the EPA and CARB had Briggs & Stratton squarely in their sights.

The regulators merely set limits for the three most common pollutants found in exhaust gases. They have not, in the past at least, presumed to dictate the technology that would be used to achieve these limits. But offroad small engines are a special case, coming on the heels of successful efforts to control emissions from automobiles, locomotives, and other sources.

The EPA points out that few garden and utility engines employ overhead valves, none have feedback control systems to monitor and control air-fuel mixtures and that no significant number have provision for exhaust-gas treatment. It seems clear that these technologies, which have made some automobile exhausts cleaner than urban air, could do something similar for small engines. CARB insists that small-engine makers adopt catalytic converters as the price of entry to the California market.

The three exhaust pollutants of concern are hydrocarbons (HC), oxides of nitrogen (NO_x, rhymes with "sox") and carbon monoxide. The first two of these combine in sunlight to form smog; CO is a poisonous, heavier-than-air gas that, even in small concentrations, adversely affects people with respiratory or heart problems.

HC is a product of incomplete combustion. Two-stroke engines, with total-loss oiling systems and fuel-contaminated exhausts, are the worst offenders. Side-valve four-cycle engines do better, but their elongated combustion chambers, necessary to house the valves, inhibit combustion. The compact combustion chambers of overhead-valve engines, symmetrical and centered over the piston, make this valve configuration the best choice HC emissions control.

With any engine, the leaner the mixture, the hotter the flame, and more complete the combustion. But, if the mixture is too lean, the engine misfires and HC emissions surge. To burn to these mixtures it is necessary to, as it were, stoke the fire by making gas flow more turbulent. One way to do this is to angle the inlet port give a spin, or swirl, to the intake charge. Another technique, used on all Briggs engines, is to incorporate a swish area between the piston crown and the roof of the chamber. As the piston nears top dead center, it "squishes" the end gases toward center of the chamber.

NO_x results from high combustion temperatures and, ironically, is reduced with rich, cooling-running mixtures. Not only do rich mixtures promote HC, they also produce large amounts of CO. The art of emission control consists of juggling compression and air-fuel ratios, ignition timing and combustion-chamber shapes to produce a minimum of NO_x, while keeping HC and CO within acceptable limits. Even the restriction imposed by the muffler has an impact on chamber temperature and NO_x formation.

Two-stroke engines emit relatively little NO_x. Scavenging, or cylinder purging, is never complete and some exhaust gases remain to quench the flame. These engines have a kind of built-in exhaust-gas recirculation.

The regulations

As things stand today, small engines are subject to EPA Phase 1 and, if sold in California, to CARB Tier 2 restrictions. Tables 9-1 and 9-2 summarize these requirements for what the regulators call "nonhandheld" engines of the type used to power mowers, generators, and the like. Exhaust-emissions limits for handheld edgers and portable tools are less stringent. HC and NO_x are treated as a single entity, since both are smog precursors. One kilowatt-hour (kW-hr) equals 1.34 horsepower-hour (hp-h).

Briggs responded by introducing overhead-valve engines with better combustion technologies and by tightening limits on air-fuel mixtures. The nonadjustable carburetor was a byproduct of this effort. The company also phased out some side-valve production, including the popular 13-cid workhorse, which it had been building at the rate of more than 5000 units a day. According to a company spokesman, exhaust emissions for 2006-model

Table 9-1
EPA Phase 1 emissions limits by model year

Displacement (cc)	2002 HC + NO$_X$ (g/kW-hr)	2003 HC + NO$_X$ (g/kW-hr)	2004 HC + NO$_X$ (g/kW-hr)	2005 HC + NO$_X$ (g/kW-hr)	2006 HC + NO$_X$ (g/kW-hr)	2007+ HC + NO$_X$ (g/kW-hr)	All years CO (g/kW-hr)
100-225	16.1	16.1	16.1	16.1	16.1	16.1	517
>225	16.6	15.0	13.4	12.1	12.1	10.0	517

Table 9-2
CARB Tier 2 emissions limits by model year

Displacement (cc)	2005 HC + NO$_X$ (g/kW-hr)	2006 HC + NO$_X$ (g/kW-hr)	2007 HC + NO$_X$ (g/kW-hr)	2008 HC + NO$_X$ (g/kW-hr)	2005+ CO (g/kW-hr)
80–225	16.1	16.1	10.0	10.0	549
>225	12.1	12.1	12.1	8.0	549

mower engines are only a quarter of the level of 1990 models. Consumers received better, more fuel-efficient products that, in the case of OHV engines, run cooler and last longer. And because these regulations point to the eventual demise of two-cycle engines, Briggs was able to move into the edger market with its 22-cc four-cycle Micro.

As additional evidence of its good faith, the company revamped its manufacturing facilities to reduce evaporative emissions from painting operations by 73% and to cut natural gas and electricity consumption.

As the century closed, EPA and CARB proposed new standards. The federal agency would reduce HC and NO$_X$ emissions from nonhandheld engines by 59% and for handheld engines 70% by the year 2027. CARB, representing the state with the nation's worst air quality, was more aggressive. Its proposal would slash small-engine emissions in half by 2008.

The EPA Phase 2 proposal would mandate converters in the distant future, in the sweet bye-and-bye when we will be living in happy retirement or learning Chinese or whatever. But no amount of tinkering with combustion chamber shapes or carburetor mixtures can achieve the level cleanliness proposed by CARB. The only option is the catalytic converter.

Catalytic converters contain small, but significant amounts of precious metals, such as platinum, palladium, and rhodium. And, for converters to live, extremely precise control of the air-fuel mixture is required. A study commissioned by B & S found that catalytic converters would boost the cost of mowers by 31%.

Briggs argues that there is no reason to single out lawn- and garden-equipment engines (that account for 80% of the company's sales) for regulation. These engines are responsible for only 3% of all smog-producing emissions (Fig. 9-1). CARB responded saying that halving emissions from these sources would be the equivalent of taking a million cars off the road. According to the agency, 2006-model lawnmowers produce 93 times more smog-forming emissions than 2006-model automobiles. To put this into context, it should be remembered that the typical mower owner uses a gallon or two of gasoline a year.

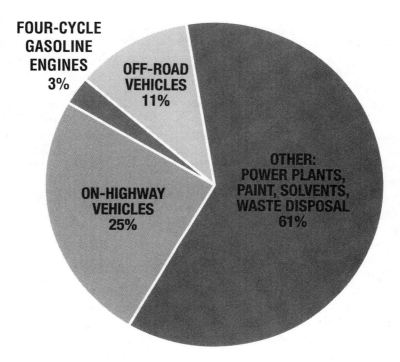

FIG. 9-1. *Contributions to smog-forming emissions as calculated by the EPA. The "four-cycle gasoline engine" category includes all nonhandheld engines, such as lawnmowers and generators.*

The company points out that catalytic converters, which operate at red heat, present fire hazards, especially when mowing high grass. Fire fighters in California testified that the hazard is real. One forest fire could generate more NO_x and HC than all the lawnmowers ever sold in California. A CARB spokesman responded that auto manufacturers had used the same fire-hazard argument when the subject of catalytic converters was first broached. "I think it's very analogous to what happened in the 70s. The arguments are all the same."

Briggs hired the prestigious National Economic Research Association to analyze the economic impact of CARB the proposal. NERA concluded that implementation of these regulations would send the company's manufacturing operations overseas and cost 22,000 jobs in 24 states. These sentiments were seconded by Thomas Savage, B & S vice-president for administration, who said that California regulators "force us to consider moving operations overseas, and this would have a tremendous impact, not only on our workers, but on our suppliers and customers."

Not everyone at Briggs was reading from the same page. In its Sept 11, 2003, SEC filing the company assured stockholders that "it does not believe" that pending CARB standards, "will have a material effect on its financial condition ... given that California represents a relatively small percentage of B & S engine sales and that increased costs will be passed on to California customers."

The threat of job losses was taken seriously by Congressman Christopher S. Bond, Rep., Missouri, where Briggs has two manufacturing plants. Bond pushed through legislation to delay the California regulations and to prohibit other states from adopting them. Bond's legislation also delayed implementation of the EPA proposals until the agency satisfied itself that catalytic converters are safe. This judgment has been made—all that's required is a heat shield—and it appears that the EPA will mandate some mix of its own and CARB limits by the time this book goes to print. A golf-ball sized catalytic converter will soon become the norm for small engines sold in California and, if the EPA so wishes, nationwide.

Compliance

To assure compliance, the EPA makes random audits of Briggs production two or three times a year. Agency employees arrive unannounced and pull engines off the line for emissions testing that can extend for 50 hours. Presumably, the agency does the same at competitor's plants.

In preparation for writing its own standards, the European Community (EC) ran emissions tests on American small engines to investigate the effectiveness of EPA Phase 1 rules.[1] Most engines were purchased anonymously to assure that they were in an as-delivered condition, without special tuning. Nine of the engines were the type used to power mowers and other nonhandheld garden equipment. Two had side-valves and six were overhead-valve models. The remaining engine was a two-stroke, which was remarkable in its way.

Engines were run on a dynamometer in accordance with EPA test schedules. Although not required by EPA, fuel consumption was measured with the OHV models easily out-performing the flatheads. Good combustion chamber design combined with the higher compression ratios translate as better fuel economy. The two-stroke came in last, as could be expected.

All nine engines produced fewer CO emissions than the 517 g/kW-hr Phase-1 limit. Combined HC/NO_x emissions depended heavily upon the HC component that, for most examples, accounted for 80% or more of the total. Because its HC emissions were an order of magnitude above the group average, the

---

[1] An Experimental Survey on the Emissions Characteristics of Small S.I. Engines for Non-Road Mobile Machinery. F.Millo, G. Cornette, and W. Meirsch, Small Engine Technology Conference and Exhibition, Pisa, Italy, Nov. 2001, Proceedings, Vol 2, pp. 609–616.

two-stroke produced ten times the legal amount of HC/NO$_x$. The four-strokes did better, but only four of the eight could meet the EPA 16.1 g/kW-hr standard for smaller engines or the 13.4 g/kW-hr standard for displacements of more than 225 cc. All those that passed had overhead valves.

These engines were legal, that is, one or more examples had been forwarded to the EPA for testing and certification prior to sale. Yet more than half failed the European tests. How could the two-stroke—that pumped out more than 160 grams of HC/NO$_x$ per kilowatt-hour—have qualified for certification?

Briggs' tests of competitive engines shows the same lack of compliance. Why tighten the rules, when the existing rules are not enforced?

And so it goes. But Briggs is determined to stay in business, regardless of emissions legislation. The company has developed and patented a catalytic converter that uses far less precious metal than previous types. Exhaust recycles through the converter in multiple passes. Briggs also has a $75 fuel injection system ready for production when the rules require it.

Other new products are on the way. One is a radical innovation known as Direct Overhead Valve, which eliminates fragile pushrods and the complication involved with driving overhead cams. These DOV engines have a single-lobe nylon camshaft in the block. Two long rocker arms bear against the lobe and drive the overhead valves. Simple and elegant.

Index

0.5 A alternators, 112
1.5 A alternators, 112–113
4 A alternators, 114–115
7.0 A alternators, 115–119
10 A alternators, 119–120

A

ac-only alternators, 113–114
air cleaners, 83
air cooling, 10–14
air filters, 82–84
air quality, 199–205
   compliance, 204–205
   regulations, 200–204
   regulatory bodies, 199–200
airvane governors, 84
alternators, 111–120
   0.5 A alternators, 112
   10 A alternators, 119–120
   1.5 A alternators, 112–113
   4 A alternators, 114–115
   7.0 A alternators, 115–119
   DC-only and AC-only alternators,
      113–114
   dual-circuit, 117
   System 3 alternators, 111
   System 4 alternators, 111
aluminum keys, 37
anti-friction ball bearings, 172
area distributors, table, 6
armature air gaps, 45–46
automatic chokes, 62, 68–70

B

ball bearings, 15, 172
batteries, 109–111
bearings, 15–16, 172–173
biodegradable cleaners, 123
blocks, 14–15
Bond, Christopher S., 204
breather assembly, 125
Briggs & Stratton starters,
   96–108
   horizontal pull starters, 97–99
   impulse starters, 102
   motors, 102–108
   sprag clutches, 96–97
   vertical pull starters, 99–102
Briggs & Stratton Walbro, 76–77
butterfly, 58
buying engines and parts, 4–7

C

California Air Resources Board
   (CARB), 199
California Air Resources Board
   (CARB) Tier 2 emissions
   limits, 201
camshaft flyweights, spring-loaded,
   10
camshafts, 146–153
   aluminum-block
      ball-bearing, side-valve engines,
      146
      camshaft bearings, 146–147

camshafts, aluminum-block *continued*
  plain-bearing, side-valve and
    OHV engines, 146
  automatic compression releases,
    151–153
  cast-iron block
    ball-bearing, horizontal-
      crankshaft engines, 148–151
    plain-bearing, horizontal-
      crankshaft engines, 148
CARB (California Air Resources
  Board), 199
CARB (California Air Resources
  Board) Tier 2 emissions limits,
  201
carbon monoxide, 199
carburetion, 17
carburetors, 57–62
  cold start, 60–62
  external adjustments, 62–63
  float-type, 65
  fuel inlet, 59–60
  one-piece Flo-Jets, 75
  service by model, 72–82
    Briggs & Stratton Walbro, 76–77
    Flo-Jet, 72–76
    Pulsa-Jet, 77–82
    Vacu-Jet, 77–82
  throttle and low-speed circuit, 58–59
  venturi and high-speed circuit,
    57–58
casting distortions, 73–74
centrifugal compression releases, 151
centrifugal governors, 84
charging systems, 109–122
  alternators, 111–120
    0.5 A, 112
    10 A, 119–120
    1.5 A, 112–113
    4 A, 114–115
    7.0 A, 115–119
    AC-only, 113–114
    DC-only, 113–114
    System 3, 111
    System 4, 111
  Nicad system, 121–122
  storage batteries, 109–111
chokes
  automatic, 62, 68–70
  failure to start, 25
  manual, 25
  plug, 61
cleaners, biodegradable, 123
cleaning, fuel system, 71–72
clip-and-spring assembly, 51
clutches
  Eaton starters, 90–93
  housing torque limits, 97
  sprag, 96–97
collet-type split keepers, 136
compression, loss of, 27–28
compression releases, 151
connecting rods, 153–158
construction, 10–17
  air cooling, 10–14
  bearings, 15–16
  blocks, 14–15
  carburetion, 17
  crankcases, 14–15
  flywheels, 16–17
  ignition, 17
  lubrication, 14
  valve mechanisms, 10
conversion, magneto-to-Magnetron,
  48–52
crankcases, 14–15
crankshafts, 174–175
cylinder bore, 166–172
  boring, 166–170
  glaze breaking, 171
  honing, 170–171
  racing engines, 171–172
cylinder heads, 129–130

D
Darlington transistor, 47
DC-only alternators, 113–114
depressor tool, 40
diagnosis, 125–129
Direct Overhead Valve, 205

disarming Eaton starters, 89
displacement, 17–19
DU bushing, 15
DU Teflon-impregnated bronze
    bushings, 172

E
Easy-Spin compression releases, 1, 151
Eaton starters, 88–96
    clutches, 90–93
    disarming, 89
    rope replacement, 89–90
    springs, 93–96
EC (European Community), 204
environmental issues, 199–205
    compliance, 204–205
    regulations, 200–204
    regulatory bodies, 199–200
Environmental Protection Agency
    (EPA), 199, 201, 202
EPA (Environmental Protection
    Agency), 199, 201, 202
Europa, 177–195
    assembly, 189–193
    disassembly, 179–188
    inspection, 188–189
    service history, 193–195
    technical description, 178–179
European Community (EC), 204
exhaust pollutants, 199–200
exhaust smoke, 31

F
factory flywheel pullers, 36
Faulkner, Daniel, 179
flanges, 139–145
    assembly, 141–143
    disassembly, 139–141
    oil pump, 144–145
    oil seal, 143–144
    oil slinger, 144–145
flathead engines, 10
floats
    Briggs & Stratton Walbro, 76–77
    setting, Flo-Jet, 72–75

float-type carburetors, 65
Flo-Jet
    crossover, 75–76
    one-piece, 74–75
    two-piece, 72–74
        casting distortion, 73–74
        float setting, 72–73
        needle and seat, 72
        throttle shaft/bearing
            replacement, 74
flooding
    carburetor, 66
    fuel, 25–26
    oil, 26
fly weights, 84
flywheels, 16–17, 34–39, 88
foam air filters, 82
footprint, vertical-shaft engines, 5
foreign partnerships, 197–198
four-stroke-cycle engines, 8, 9
fuel
    flooding, 25–26
    no delivery, 64–65
    starvation, 26–27
fuel inlets, 59–60
fuel pumps, 82–84
fuel system, 57–86
    carburetor service by model,
        72–82
        Briggs & Stratton Walbro, 76–77
        Flo-Jet, 72–76
        Pulsa-Jet, 77–82
        Vacu-Jet, 77–82
    carburetors, 57–62
        cold start, 60–62
        fuel inlet, 59–60
        throttle and low-speed circuit,
            58–59
        venturi and high-speed circuit,
            57–58
    cleaning, 71–72
    external adjustments, 62–63
    fuel pumps, 82–84
    governors, 84–86
    installation, 67–71

fuel system, installation *continued*
   automatic chokes, 68–70
   Pulsa-Jet tank diaphragms, 70–71
  removal, 67–71
   automatic chokes, 68–70
   Pulsa-Jet tank diaphragms,
    70–71
  repair, 71–72
  troubleshooting, 64–67
   carburetor floods, 66
   engine refuses to idle, 66–67
   engine runs lean at full throttle,
    65
   engine runs rich, 65–66
   no fuel delivery, 64–65

G
Gerotor pump, 145
"good-will policy," 7
governors, 84–86
  airvane, 84
  centrifugal, 84
  mechanical, 86
Greased Lightening biodegradable
  cleaners, 123
grinding valves, 139

H
HC (hydrocarbons), 199–200
head fins, 12
hex-nut, 35
high-speed nozzle, 58
holding fixture, vertical-shaft
  engines, 168–169
horizontal pull starters, 97–99
horizontal-shaft engines, 5
horsepower (hp), 17–19
hydrocarbons (HC), 199–200
hydrometers, 110

I
identification of engines, 2–3
idling
  engine refuses, 66–67
  failure, 30

ignition, 17
  failure to start, 24–25
  loss of, 53
ignition systems, 33–55
  armature air gaps, 45–46
  flywheels, 34–39
  magnetos, 39–45
  Magnetrons, 46–52
   magneto-to-Magnetron
    conversion, 48–52
   replacing existing switch
    module, 52
   service, 48
  safety interlocks, 53–55
  shutdown switches, 53
  spark plugs, 33–34
ignition testers, 21
I-head engines, 10
impulse starters, 102
initial mixture screws, 62–63
inlet seats, 76
installation, 67–71
  automatic chokes, 68–70
  Pulsa-Jet tank diaphragms,
   70–71
  valves, 136–137
interlocks, 55
iron blocks, 15

K
keys, 35
keyway, 38
kickback, 30

L
lash adjustment, 137–138
L-head engines, 10, 12
Loctite Natural Blue biodegradable
  cleaners, 123
lubrication, 14

M
Magna-Matics, 43, 44
magnetos, 39–45
  magneto-side seals, 173

magneto-to-Magnetron conversion, 48–52
Magnetrons, 46–52
  magneto-to-Magnetron conversion, 48–52
  replacing existing switch module, 52
  service, 48
main air bleed, 57
main air jet, 57
main-bearing loads maximums, 16
manometers, 125–129
manual chokes, 25
mechanical governors, 86
mixture-adjustment screw, 80
model number codes, 3
motors, starter, 102–108

N
National Economic Research Association (NERA), 203
NC (normally closed) interlocks, 55
needles, Flo-Jet
  one-piece, 74
  two-piece, 72
needle-valve assemblies
  Pulsa-Jet, 80–81
  Vacu-Jet, 80–81
NERA (National Economic Research Association), 203
Nicad system, 109, 121–122
NO (normally open) interlocks, 55
nomenclature, 10–17
  air cooling, 10–14
  bearings, 15–16
  blocks, 14–15
  carburetion, 17
  crankcases, 14–15
  flywheels, 16–17
  ignition, 17
  lubrication, 14
  valve mechanisms, 10
normally closed (NC) interlocks, 55
normally open (NO) interlocks, 55

NOx (oxides of nitrogen), 199–200
nylon starter cord, recoil starter, 87

O
OEMs (original equipment manufacturers), 2
oil
  consumption factors, 127
  flooding, 26
  pump, 144–145
  seal, 143–144
  slinger, 144–145
oil-control rings, 162
one-piece Flo-Jets, 75
operating cycle, 8–10
original equipment manufacturers (OEMs), 2
overhead-valve engines, 10, 158
overhead-valve mechanism, 12
oxides of nitrogen (NOx), 199–200

P
paper air filters, 82
parts, buying, 7
pickup tubes, 81
piston rings, 162
pistons, 158–161
plain bearings, 172
plug chokes, 61
PN 19051 ignition testers, 21
PN 19368 ignition testers, 21
pollutants, exhaust, 199–200
power, loss of, 29
power curve, 17
power output factors, 126
pressed steel or aluminum housing, recoil starter, 87
primary idle port, 58
primer, 25
pulley, recoil starter, 87
Pulsa-Jet, 77–82
  needle-valve assembly, 80–81
  pickup tubes, 81
  pump diaphragm, 82
  tank diaphragms, 70–71

pump diaphragms, Pulsa-Jet, 82
pushrods, 28

R
radius-face rings, 163
recoil starters, 87
rectifier tester, 122
repair, 123–175
  camshafts, 146–153
    aluminum-block engines,
      146–147
    automatic compression releases,
      151–153
    cast-iron block engines, 148–151
  connecting rods, 153–158
  crankshafts, 174–175
  cylinder bore, 166–172
    boring, 166–170
    glaze breaking, 171
    honing, 170–171
    racing engines, 171–172
  cylinder heads, 129–130
  diagnosis, 125–129
  flanges, 139–145
    assembly, 141–143
    disassembly, 139–141
    oil pump, 144–145
    oil seal, 143–144
    oil slinger, 144–145
  fuel system, 71–72
  magneto-side seals, 173
  main bearings, 172–173
  pistons, 158–161
  resources, 123–124
  rings, 162–165
  valves, 130–139
    grinding, 139
    guides, 139
    installation, 136–137
    lash adjustment, 137–138
    removal, 134–136
    seats, 139
replacement
  ropes, 89–90
  switch modules, 52

throttle shaft/bearing, Flo-Jet, 74
Repower Project, 7
rewind starters, troubleshooting,
  87–88
rheostat OMC, 110
ring grooves, pistons, 158
rings, 162–165
  cross-sections, 162
  oil-control, 162
  piston, 162
  radius-face, 163
  scraper, 162
ropes
  broken, 88
  hard to pull, 88
  replacement, 89–90
rotations per minute (rpm), 17–19
rpm (rotations per minute), 17–19

S
safety interlocks, 53–55
Savage, Thomas, 203
scraper rings, 162
scuffing pistons, 158
seals
  magneto-side, 173
  oil, 143–144
seats, Flo-Jet
  one-piece, 74
  two-piece, 72
service
  carburetors, 72–82
    Briggs & Stratton Walbro, 76–77
    Flo-Jet, 72–76
    Pulsa-Jet, 77–82
    Vacu-Jet, 77–82
  Magnetrons, 48
sheave, recoil starter, 87
short-circuit test, 114
shutdown, 30, 53
side-pull starters, Eaton-pattern, 88–96
  clutches, 90–93
  disarming, 89
  rope replacement, 89–90
  springs, 93–96

side-valve engines, 10
  torque limits, 132
  torque sequences, 131
  valve lash, 138
single-cylinder engines, 38
slingers, 14
smog-forming emissions, 203
spark plugs, 33–34
splash lubrication, 14
sprag clutches, 96–97
spring-loaded camshaft flyweights, 10
springs, 93–96
starter clutch, 35
starters, 87–108
  binding, 30–31
  Briggs & Stratton starters, 96–108
    horizontal pull starters, 97–99
    impulse starters, 102
    motors, 102–108
    sprag clutches, 96–97
    vertical pull starters, 99–102
  Eaton-pattern, side-pull starters, 88–96
    clutches, 90–93
    disarming, 89
    rope replacement, 89–90
    sheaves, 93–96
    springs, 93–96
  rewind starters, 87–88
starting
  cold start, 60–62
  failure, 24–27
    fuel flooding, 25–26
    ignition, 24–25
    no fuel, 26–27
    oil flooding, 26
    primer, 25
steady draw current ratings, 104
steel keys, 37
strap wrench, 36
switch modules, 52
switches, shutdown, 53
System 3 alternators, 111
System 4 alternators, 111

T
tank diaphragms, Pulsa-Jet, 70–71
taper, 155
throttles, 5, 58–59, 65, 74
thrust faces, pistons, 158
torque, 17–19, 131, 132
troubleshooting, 21–32
  engine runs briefly and quits, 29
  excessive vibration, 31–32
  exhaust smoke, 31
  failure to idle, 30
  failure to shutdown, 30
  failure to start, 24–27
    choke, 25
    fuel flooding, 25–26
    ignition, 24–25
    no fuel, 26–27
    oil flooding, 26
    primer, 25
  fuel system, 64–67
    carburetor floods, 66
    engine refuses to idle, 66–67
    engine runs lean at full throttle, 65
    engine runs rich, 65–66
    no fuel delivery, 64–65
  kickback, 30
  loss of compression, 27–28
  loss of power, 29
  rewind starters, 87–88
  starter binds, 30–31
two-piece Flo-Jet, 63

U
U-tube manometers, 128

V
Vacu-Jet, 77–82
  needle-valve assembly, 80–81
  pickup tubes, 81
valve lash, side-valve engines, 138
valve mechanisms, 10
valve-in-head engines, 10

valves, 130–139
  grinding, 139
  guides, 139
  installation, 136–137
  lash adjustment, 137–138
  loss of compression, 28
  needle and seat, 59
  removal, 134–136
  seats, 139
venturis, 57–58
vertical pull starters, 99–102

vertical-shaft engines
  footprint, 5
  holding fixture, 168–169
  work stand, 124
vibration, excessive, 31–32

W
Wisconsin Robin W1-145 V
    engines, 11
work stand, vertical-shaft engines, 124